Physical Principles
of Audiology

Medical Physics Handbooks
Other books in the series

Series Editor: **Professor J M A Lenihan**
Department of Clinical Physics and Bio-Engineering
West of Scotland Health Boards, Glasgow

Medical Physics Handbooks 3

Physical Principles of Audiology

P M Haughton
Medical Physics Department,
Hull Royal Infirmary

Adam Hilger Ltd, Bristol
in collaboration with the
Hospital Physicists' Association

187620 610.28
H 371

© 1980 P M Haughton

British Library Cataloguing in Publication Data

Haughton, P M
    Physical Principles of audiology.
    (Medical physics handbooks; 3)
    ISSN 0143-0203
    1. Audiology
    I. Title II. Series
    617.8'9   RF294

    ISBN 0-85274-502-8

Published by Adam Hilger Ltd,
Techno House, Redcliffe Way, Bristol BS1 6NX

The Adam Hilger book-publishing imprint is owned by The Institute of Physics

Filmset by The Universities Press (Belfast) Ltd
and printed in Great Britain by The Pitman Press, Lower Bristol Road, Bath BA2 3BL

Contents

Preface

This book provides a concise introduction to major topics in audiology. It is intended primarily for the reader who has had some training in physics or an allied discipline, but a previous knowledge of medicine or biology is not expected.

Audiology is the product of many different disciplines rather than a self-contained science, and a short monograph such as this cannot provide more than an outline of so diverse a subject. The aim has therefore been to give the reader an appreciation of the fundamentals of those topics which seem particularly important. To this end, it has been necessary to omit the kind of detail which would be of interest mainly to specialists. An exception to this approach has been made in Chapters 5 and 6, which include technical information on the calibration of audiometers and the clinical application of acoustic impedance measurement. This information has been included to cater for the needs of the growing number of physicists, audiological scientists and technicians responsible for the supervision of hospital audiology services.

I am especially grateful to Mr R G Williams, TD, FRCS, for his constant help and encouragement throughout the preparation of this book. I also wish to thank Mr D Thompson and the staff of the Medical Library, Hull Royal Infirmary, for their help in obtaining the many books and journals which had to be consulted.

P M Haughton
Hull, February 1979

1 Acoustics

1.1 Introduction

A vibrating body immersed in an elastic medium radiates sound. Acoustics is the study of this radiated energy and the behaviour of the transmitting medium. The word 'sound' may be used to denote any mechanical disturbance propagated in the medium, whether audible or not, though its use is often restricted to frequencies in the audible range. Disturbances at higher frequencies are described as *ultrasonic*, while those below the audible range are *infrasonic*. For most purposes the medium is treated as a continuum and a *particle* of it is an infinitesimal volume which is nevertheless imagined to retain the properties of the bulk material. Alternatively, a particle may be defined as a volume which is large compared with molecular dimensions, but small compared with the wavelength of the prevailing sound.

1.2 Transmission of Sound in Fluids

In what follows, only media in which the shear modulus of elasticity is negligible will be considered. In such materials sound waves are periodic fluctuations in density corresponding to local variations in pressure. Gravity and capillary waves, for example, are not sound waves because they are not associated with changes in density. In viscoelastic liquids and in true solids sound may also be transmitted as shear and bending waves. This mode of propagation will not be considered.

1.2.1 Some definitions

Static pressure (p_0) at a point in the medium is the pressure which would exist in the absence of sound.

Sound pressure (p) is the alternating component of the pressure at a point in a sound field.

Condensation (*s*) is the ratio of the increment of density $\delta\rho$ to the density ρ_0 of the undisturbed medium. Thus $s = \delta\rho/\rho$ and $\rho = \rho_0 (1+s)$, where ρ_0 and ρ are the densities at pressures p_0 and $p_0 + p$, respectively.

Bulk modulus of elasticity (κ) is the ratio of an increment of pressure δP to the associated increment of volume strain $-\delta V/V$, where $P = p_0 + p$. Thus

$$\kappa = -V\frac{\delta P}{\delta V} = \frac{p}{s}.$$

Particle velocity (*u*) is the alternating component of the velocity of movement of the medium at a point in a sound field.

Velocity potential (ϕ) is a scalar potential function defined for an isotropic medium at rest or in steady irrotational flow, by the relationship $u_x = -(\partial\phi/\partial x)$, where u_x is the component of the particle velocity in the *x* direction.

Intensity (*I*) of a sound in a specified direction is the average rate of flow of acoustic energy through a unit area normal to that direction.

Energy density (*E*) is the acoustic energy per unit volume at a point in the medium.

1.2.2 Wave motion in fluids

Fluids as defined above are non-dispersive; the velocity at which a disturbance is propagated is independent of the frequency. The speed of sound (*c*) depends on the bulk modulus and the density of the medium,

$$c^2 = \kappa/\rho_0 . \tag{1.1}$$

For waves in a gas (usually air) the appropriate elastic constant is, except in special circumstances, the adiabatic modulus, so that $\kappa = \gamma p_0$, where γ is the ratio of the principal specific heats. For small amplitudes, wave propagation may be expressed in terms of the condensation,

$$\frac{\partial^2 s}{\partial t^2} = c^2 \nabla^2 s, \tag{1.2}$$

where

$$\nabla^2 s = \frac{\partial^2 s}{\partial x^2} + \frac{\partial^2 s}{\partial y^2} + \frac{\partial^2 s}{\partial z^2}$$

in cartesian coordinates.

Other important relationships are:

$$\frac{\partial u_x}{\partial t} = -c^2 \frac{\partial s}{\partial x}, \tag{1.3}$$

and the corresponding equations for the velocity components u_y and u_z; also

$$\frac{\partial \phi}{\partial t} = c^2 s \tag{1.4}$$

and, since $c^2 = \kappa/\rho_0$ and $s = p/\kappa$,

$$\frac{\partial \phi}{\partial t} = \frac{p}{\rho_0}. \tag{1.5}$$

In terms of velocity potential the wave equation is

$$\frac{\partial^2 \phi}{\partial t^2} = c^2 \nabla^2 \phi. \tag{1.6}$$

1.2.3 Plane progressive waves

For a wave travelling in the positive x direction, the solution of equation (1.6) has the form $\phi = f(ct - x)$. For a simple harmonic wave† let

$$\phi = A \exp[i(\omega t - kx)],$$

where ω is the angular frequency and k is the *phase change coefficient* or *wavelength constant* $(k = \omega/c = 2\pi/\lambda)$.

By equation (1.5)

$$p = \rho_0(\partial \phi/\partial t) = i\omega\rho_0 A \exp[i(\omega t - kx)]$$

and

$$u = u_x = -(\partial \phi/\partial x) = ikA \exp[i(\omega t - kx)].$$

---

† The use of bold type to distinguish complex quantities from their magnitudes is not adopted here. Magnitudes will be indicated by the use of appropriate subscripts or magnitude signs.

Hence

$$p = \omega \rho_0 u / k = \rho_0 c u . \tag{1.7}$$

The particle displacement ξ is given by

$$\xi = -\frac{iu}{\omega} = \frac{-ip}{\omega \rho_0 c} , \tag{1.8}$$

so that

$$\xi_{rms} = \frac{u_{rms}}{\omega} = \frac{p_{rms}}{\omega \rho_0 c} . \tag{1.9}$$

The energy E per unit volume of the medium is the sum of the potential energy due to stored elastic energy and the kinetic energy associated with the particle velocity. Although the average potential energy is equal to the average kinetic energy, the sum of their instantaneous values is not constant, so that E is an alternating quantity.

$$E = \tfrac{1}{2} p^2 / \kappa + \tfrac{1}{2} \rho_0 u^2 . \tag{1.10}$$

For plane waves p and u are related by equation (1.7) so that the time-average energy density is

$$E_{av} = \rho_0 u_{rms}^2 = \frac{p_{rms}^2}{\rho_0 c^2} = \omega^2 \rho_0 \xi_{rms}^2 . \tag{1.11}$$

In unit time the energy propagated through unit area normal to the direction of propagation is contained in a volume c. Hence the intensity is given by

$$I = E_{av} c = \rho_0 c u_{rms}^2 . \tag{1.12}$$

1.2.4 Spherical waves
For spherical symmetry

$$\nabla^2 \phi = (1/r) \frac{\partial^2 (r\phi)}{\partial r^2} .$$

The solution to equation (1.6) for a divergent spherical wave has the form

$$\phi = (1/r) f(ct - r) .$$

For a simple harmonic wave,

$$\phi = (A/r) \exp [i(\omega t - kr)] .$$

Thus the particle velocity is given by

$$u = -\frac{\partial \phi}{\partial r} = \frac{A}{r}(1/r + ik)\exp[i(\omega t - kr)]. \qquad (1.13)$$

The pressure is

$$p = \rho_0 \frac{\partial \phi}{\partial t} = \frac{A}{r}\rho_0 ck \exp[i(\omega t - kr)]. \qquad (1.14)$$

The peak volume flux, Q, at the surface of a spherical source whose radius a is small ($ka \ll 1$), is given by $Q = 4\pi a^2 u_p$, where u_p is the peak velocity of the surface of the source. Thus by equation (1.13), $Q = 4\pi A$, and in the above equations A can be replaced by $Q/4\pi$. The flux Q is called the *strength* of the source. The rms pressure is therefore given by

$$p_{rms} = \frac{\rho_0 ckQ}{4\sqrt{2}\,\pi r}. \qquad (1.15)$$

Thus the sound pressure is inversely proportional to the distance from the source. This simple law does not apply to the particle velocity in a divergent field.

The intensity of the wave is obtained by averaging the product of the sound pressure and the component of the particle velocity that is in phase with the pressure. Thus

$$I = \frac{\rho_0 ck^2 Q^2}{32\pi^2 r^2}. \qquad (1.16)$$

The average power P of the source is the intensity multiplied by the surface area of a sphere of radius r, so that

$$P = \frac{\rho_0 ck^2 Q^2}{8\pi}. \qquad (1.17)$$

The application of equations (1.15) and (1.16) is not restricted to spherical sources; these equations are also valid for any simple source† of strength Q whose size is small compared with the wavelength and to the distance r at which P and I are required.

---

† A simple source is one that radiates uniformly in all directions. At a distance r any source is equivalent to a simple source if its free-field radiation is uniform in all directions.

1.3 Acoustic and Mechanical Impedance

The *acoustic impedance* (Z_a) is the ratio of the sound pressure to the volume velocity U in the medium. The latter is the rate of flow of fluid through a specified area S. Thus

$$U = \int_S u \cdot dS .$$

In what follows, S will be an area over which the particle velocity is constant and normal to the surface at all points. Thus $U = uS$, and

$$Z_a = p/uS . \qquad (1.18)$$

The *specific acoustic impedance* (Z_s) is the corresponding relationship between pressure and particle velocity:

$$Z_s = p/u . \qquad (1.19)$$

Similarly, the *mechanical impedance* (Z_m) is the ratio of the force f to the velocity u of a mechanical system, where u and f are in the same direction:

$$Z_m = f/u . \qquad (1.20)$$

If the mechanical system is a vibrating piston of area S operating against a sound pressure p, the alternating force on it is pS. The piston and the adjacent fluid medium must move with the same velocity, so that equations (1.18) and (1.20) combine to give

$$Z_m = Z_a S^2 . \qquad (1.21)$$

The quantity Z_m is also known as the *radiation impedance* of the piston.

1.3.1 Acoustic impedance in plane and spherical waves
Equation (1.7) shows that, for plane waves, the specific acoustic impedance is given by

$$Z_s = \rho_0 c. \qquad (1.22)$$

The quantity $\rho_0 c$ is a real number, that is, the particle velocity and sound pressure are in phase. This impedance depends only on the properties of the medium and is called the *characteristic* impedance (or resistance) of the medium. For spherical waves, equations (1.13) and

(1.14) give

$$Z_s = \frac{\rho_0 ckr(kr + i)}{1 + k^2 r^2}.$$ (1.23)

The spherical wave impedance approximates to that of a plane wave when kr is large compared with unity $(r \gg \lambda/2\pi)$.

1.4   Electrical Analogues of Acoustic and Mechanical Elements

There is a direct analogy between the components of electrical, mechanical and acoustic systems. It is often helpful to represent acoustic or mechanical arrangements by their equivalent electrical circuits because familiarity with the latter aids analysis. Moreover, the performance of complex acoustic systems can be determined by the actual construction of the analogous electrical circuit (for example, Møller (1960) has used this technique in measurements of the impedance of human ears).

Acoustic pressure and volume velocity are analogous to electrical potential difference and current, and their ratios—acoustic and electrical impedance—are equivalent. The corresponding mechanical quantities are force, velocity and mechanical impedance. Impedance may be written in the form $Z = R + iX$, where R and X, which are real quantities, are called resistance and reactance in all three systems.

The rate of production of heat, $I^2 R$, in an electrical resistor is analogous to the heat dissipation $u^2 R$ in a mechanical resistance. In the latter, this loss is associated with friction. Fluid friction (viscosity) rather than solid friction is involved because the definition of R requires that the force be proportional to velocity—symbolic representations of resistive elements are sometimes misleading in this respect. Acoustic resistance is not necessarily associated with the production of heat. For example, in a spherical wave, $U^2 R$ is the acoustic power radiated through an area S of the medium. Thus intensity in equation (1.16) is the product of the mean square particle velocity and the real part of the specific impedance obtained from equations (1.13) and (1.23).

Reactance in an electrical circuit derives from inductive and capacitive components ωL and $-1/\omega C$. In the equivalent mechanical system the reactance of an oscillating mass is ωm and the reactance of an elastic element is $-1/\omega k$, where k is the *compliance* of the element.

Compliance, which is the reciprocal of *stiffness*, is the extension of the element produced by the application of unit force. The acoustic equivalents are *acoustic inertance* (M) and *acoustic compliance* (C). Inertance is the pressure per unit rate of change of volume velocity and compliance the volume displacement per unit pressure. The use of these terms is illustrated by considering the equivalent circuit of a Helmholtz resonator.

1.4.1 Helmholtz resonator
This device consists of a cavity of volume V, with a small opening formed by a short tube (the neck) of effective length L and cross sectional area S (figure 1.1). The air in the neck provides a vibrating mass coupled to an elastic element, the air in the cavity. The mass is $\rho_0 SL$, which gives an acoustic reactance $\omega \rho_0 SL/S^2$ (see equation (1.21)). The inertance of the neck is therefore

$$M = \rho_0 L/S. \tag{1.24}$$

The acoustic compliance of the cavity is

$$C = V/\kappa, \tag{1.25}$$

where κ is the bulk modulus of the air in the cavity. The acoustic reactance of the combination is

$$X_a = \omega M - \frac{1}{\omega C} = \frac{\omega \rho_0 L}{S} - \frac{\kappa}{\omega V}.$$

At resonance this reactance is zero so that the resonant frequency is

$$f = \frac{1}{2\pi\sqrt{MC}} = \frac{\kappa S}{2\pi\sqrt{\rho_0 LV}}. \tag{1.26}$$

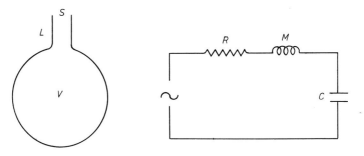

Figure 1.1 The Helmholtz resonator and its equivalent circuit.

The oscillator is damped by radiation of acoustic energy from the neck and viscous forces. These combine to produce the resistive element shown in figure 1.1.

1.4.2   Acoustic admittance

It is sometimes advantageous to use, instead of impedance, the reciprocal quantity, *admittance* (Y). Thus

$$1/Z_a = Y_a = G_a + iB_a, \qquad (1.27)$$

where G_a is the acoustic *conductance* and B_a is the acoustic *susceptance*.

1.5   Units of Impedance and Admittance

The impedances defined in §1.3 have the following MKS units: Z_m ($kg\,s^{-1}$ or $N\,s\,m^{-1}$); Z_a ($kg\,s^{-1}\,m^{-4}$ or $N\,s\,m^{-5}$); Z_s ($kg\,s^{-1}\,m^{-2}$ or $N\,s\,m^{-3}$). The corresponding CGS units are sometimes called mechanical or acoustic ohms. In the MKS system the unit of specific acoustic impedance (Z_s) is sometimes called the rayl†, but the term ohm is still retained for the other two quantities (Z_m and Z_a). To avoid confusion, the prefix MKS is often employed (MKS ohm). In many publications the term mho is used to denote the CGS unit of acoustic admittance.

Another frequently used unit is one of *equivalent volume*. The equivalent volume of an acoustic impedance Z_a is the volume of a hard-walled cavity containing air at standard temperature and pressure, and having an impedance whose magnitude is equal to $|Z_a|$. Thus the equivalent volume of an impedance Z_a or admittance Y_a is a volume V, expressed in millilitres, such that

$$|Z_a| = \frac{1}{|Y_a|} = \frac{\kappa}{\omega V} = \frac{\rho_0 c^2}{2\pi f V}. \qquad (1.28)$$

Medical instruments which really measure the admittance $|Y_a|$ of the ear often have meters indicating 'compliance' in millilitres. These instruments operate at low frequencies, for which the conductive part of the admittance is negligible. Under these conditions the reading on the meter may be interpreted either as the volume of a cavity whose

---

† The name rayl appears in Kinsler and Frey (1962) but is not included in the HMSO publication on the SI system (Page and Vigoureux 1977).

admittance is equal to the admittance of the ear, or as the volume of a cavity whose compliance is equal to that of the ear.

1.6 Reflection and Transmission at the Boundary between Two Media

Reflection occurs when a sound wave encounters a discontinuity in the acoustic impedance of the transmitting system. Such discontinuities are present at the boundaries between different fluids and in places where the area through which the sound is propagated undergoes an abrupt change. An example of the latter is the exit of a pipe.

As an illustration, consider the reflection of a plane progressive wave propagated perpendicularly to the boundary between two media whose characteristic impedances are Z_1 and Z_2 (figure 1.2). If the extent of the surface at the boundary is large compared with the wavelength, the incident wave is divided into a normally reflected component in medium 1 and a transmitted component in medium 2. The sound pressures in the incident, reflected and transmitted waves may be represented respectively by the equations

$$p_i = A_i \exp\left[i(\omega t - k_1 x)\right], \qquad p_r = A_r \exp\left[i(\omega t + k_1 x)\right]$$

and

$$p_t = A_t \exp\left[i(\omega t - k_2 x)\right].$$

The particle velocity and pressure (and other physical quantities) are continuous at the boundary. Therefore, at the boundary,

$$p_i + p_r = p_t \tag{1.29}$$

and

$$A_i + A_r = A_t. \tag{1.30}$$

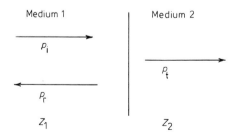

Figure 1.2 Reflection at a boundary between two media.

Similarly, for the corresponding particle velocities at the boundary,

$$u_i + u_r = u_t, \tag{1.31}$$

but

$$u_i = p_i/Z_1, \qquad u_t = p_t/Z_2 \tag{1.32a,b}$$

and for the *reflected* wave,

$$u_r = -p_r/Z_1 . \tag{1.32c}$$

From these equations

$$\frac{p_r}{p_i} = K = \frac{A_r}{A_i} = \frac{Z_2 - Z_1}{Z_2 + Z_1}, \tag{1.33}$$

where K is the complex *pressure reflection coefficient*. The *sound reflection coefficient* (α_R) is the ratio of the reflected to incident intensity. Thus

$$\alpha_R = \frac{A_r^2}{A_i^2} = K^2 = \frac{|Z_2 - Z_1|^2}{|Z_2 + Z_1|^2} . \tag{1.34}$$

Similarly, $1 - \alpha_R$ is the sound transmission coefficient (α_T).

For media of infinite extent, Z can be replaced by ρc (equation (1.22)), so that

$$\alpha_R = \frac{(r-1)^2}{(r+1)^2} \tag{1.35}$$

where $r = \rho_2 c_2 / \rho_1 c_1$, the ratio of the characteristic impedances of the media. The condition for optimum transmission (no reflection) is that these impedances are equal.

An important reciprocal property of the transmission process is that the transmission coefficient for a wave going from medium 1 to medium 2 is the same as that for a wave travelling in the opposite direction. This is a particular instance of a general reciprocity principle applicable to acoustic phenomena.

1.7 Acoustic Measurements

The quantities of interest are chiefly pressure, intensity and related quantities in the sound field of a given source. It is usually convenient to express the magnitudes of these quantities in decibel units.

Table 1.1 Physical constants relating to the transmission of
sound in air.

Temperature	20 °C
Pressure	760 mm mercury ($1 \cdot 013 \times 10^5$ Pa)
Density of dry air, ρ_0	$1 \cdot 205$ kg m^{-3}
Ratio of specific heats, γ	$1 \cdot 403$
Adiabatic bulk modulus, γp_0	$1 \cdot 421 \times 10^5$ N m^{-2}
Velocity of sound, c	$343 \cdot 5$ m s^{-1}
Characteristic impedance, $\rho_0 c$	414 N s m^{-3}
Viscosity, η	$1 \cdot 84 \times 10^{-5}$ N s m^{-2}

1.7.1 The decibel

The bel, and hence the decibel (dB), are defined fundamentally in
terms of the sound power of a source: two powers P_1 and P_2 are
separated by n decibels when

$$n = 10 \log_{10} (P_1/P_2).$$

It frequently happens that the ratios of the sound pressures, particle
velocities or analogous electrical quantities in the measuring instrument are proportional to the square root of the corresponding power
ratio. Under these circumstances,

$$n = 20 \log_{10} (p_1/p_2), \quad \text{etc.}$$

The term 'level' (e.g. pressure level, intensity level, etc) is used in
conjunction with a quantity related to power to denote the magnitude
of this quantity expressed in dB relative to a specified reference.
Unless otherwise stated, the reference sound power for measurements
in air is 10^{-12} W and the reference pressure is 2×10^{-5} N m^{-2} (20 μPa),
rms. In some circumstances, such as in the measurement of acoustic
noise and in audiometry, it is desirable to choose a reference pressure
which varies with frequency. Thus a sound pressure level (SPL) expressed in dB(A) is a level relative to a reference pressure which varies
with frequency according to the A-weighting (see table 1.2). The
terms *sensation level* or *hearing level* are used in audiometry to denote
an intensity level relative to the threshold of hearing (abbreviated to
dB SL).

Table 1.2 IEC specifications for A, B and C weighting networks. Reproduced by kind permission of the International Electrotechnical Commission, to which copyright belongs.

Frequency (Hz)	Curve A (dB)	Curve B (dB)	Curve C (dB)	Tolerance limits (dB) for curves A, B and C	
10	−70·4	−38·2	−14·3	3	−∞
12·5	−63·4	−33·2	−11·2	3·0	−∞
16	−56·7	−28·5	−8·5	3·0	−∞
20	−50·5	−24·2	−6·2	3·0	−3·0
25	−44·7	−20·4	−4·4	2·0	−2·0
31·5	−39·4	−17·1	−3·0	1·5	−1·5
40	−34·6	−14·2	−2·0	1·5	−1·5
50	−30·2	−11·6	−1·3	1·5	−1·5
63	−26·2	−9·3	−0·8	1·5	−1·5
80	−22·5	−7·4	−0·5	1·5	−1·5
100	−19·1	−5.6	−0·3	1·0	−1·0
125	−16·1	−4·2	−0·2	1·0	−1·0
160	−13·4	−3·0	−0·1	1·0	−1·0
200	−10·9	−2·0	0	1·0	−1·0
250	−8·6	−1·3	0	1·0	−1·0
315	−6·6	−0·8	0	1·0	−1·0
400	−4·8	−0·5	0	1·0	−1·0
500	−3·2	−0·3	0	1·0	−1·0
630	−1·9	−0·1	0	1·0	−1·0
800	−0·8	0	0	1·0	−1·0
1000	0	0	0	1·0	−1·0
1250	0·6	0	0	1·0	−1·0
1600	1·0	0	−0·1	1·0	−1·0
2000	1·2	−0·1	−0·2	1·0	−1·0
2500	1·3	−0·2	−0·3	1·0	−1·0
3150	1·2	−0·4	−0·5	1·0	−1·0
4000	1·0	−0·7	−0·8	1·0	−1·0
5000	0·5	−1·2	−1·3	1·5	−1·5
6300	−0·1	−1·9	−2·0	1·5	−2·0
8000	−1·1	−2·9	−3·0	1·5	−3·0
10 000	−2·5	−4·3	−4·4	2·0	−4·0
12 500	−4·3	−6·1	−6·2	3·0	−6·0
16 000	−6·6	−8·4	−8·5	3·0	−∞
20 000	−9·3	−11·1	−11·2	3·0	−∞

1.7.2 Types of sound field

A *free field*† is a sound field in a medium so extensive that the effects of its boundaries are negligible throughout the region of interest. For a given source, the region in which free-field conditions exist is called the *direct field* of the source. The direct field has two parts, namely, the *near field* in which particle velocity and pressure are not in phase, and the *far field* in which they are in phase. In an enclosure such as a room, the superposition of reflected waves from the walls produces a *reverberant field*. A *diffuse field* exists when reflections are sufficiently numerous and random that the energy density is uniform.

When measurements relating to a small source, such as a loudspeaker, are made indoors under conditions which are not anechoic, there will be a region, the direct field, in which the sound pressure is inversely proportional to the distance from the source (equation (1.15)). Close to the source, in its near field, the particle velocity is not inversely proportional to this distance, although the pressure may be if the dimensions of the source are sufficiently small. As the distance from the source is increased the reverberant or diffuse field is entered wherein the sound pressure and particle velocity are constant and in phase.

1.7.3 Microphones

Microphones are designed to respond either to sound pressure (pressure microphones) or to its space derivative (pressure gradient microphones). Carbon, crystal, condenser and moving coil ('dynamic') microphones are all pressure-operated. In these microphones the sensing element is a flexible diaphragm. The voltage generated in response to an alternating pressure at the diaphragm is proportional to the displacement or velocity of the diaphragm, depending on the type of microphone. The mechanical impedance at the diaphragm has therefore to complement the method of voltage generation in the frequency range for which the microphone is required to operate. When the output depends on displacement, as in the condenser and crystal microphones, the mechanical system is stiffness-controlled. The output of the dynamic microphone is proportional to the velocity of the moving coil and the system has to be resistance-controlled. This is achieved by constructing what amounts to a band-pass filter comprising two resonators, one mechanical and the other acoustic.

† In audiometry the term 'free field' is often used loosely to denote sound produced by a loudspeaker.

The ribbon microphone is a pressure gradient microphone. It has a small metal ribbon suspended between the poles of a magnet. The force on the ribbon is proportional to the first derivative of the sound pressure and its movement is governed by its effective mass, m. The output voltage is proportional to the velocity of the ribbon, given by

$$V_{rms} = \frac{1}{\omega m}\left(\frac{\partial p}{\partial x}\right)_{rms}.$$

When the acoustic impedance in the sound field is resistive (equal to $\rho_0 c$),

$$\left(\frac{\partial p}{\partial x}\right)_{rms} = \frac{\omega p_{rms}}{c},$$

which makes the pressure response of the microphone independent of frequency. An important feature of the ribbon microphone is that its response is highly directional.

When an object is placed in a sound field, the field is disturbed to a greater or lesser extent depending on the size of the object. This phenomenon is known as *diffraction* and it includes reflection and refraction as special cases. As a consequence of diffraction the pressure registered by a microphone differs from the pressure which would exist at the same point if the microphone were removed. The *sensitivity* of a microphone is the ratio of its electrical output (usually the open-circuit EMF) to the pressure existing under specified conditions. Thus the *pressure sensitivity* is defined with respect to the pressure that actually exists at the microphone, whereas the *free-field sensitivity* relates to the pressure in a previously unobstructed plane progressive wave. The free-field sensitivity depends on the angle between the microphone and the direction of the incident wave. The rms value of the sensitivities for all angles of incidence is the *random-incidence sensitivity*.

When the dimensions of the microphone are small compared with the wavelength, difffraction is negligible. Thus at low frequencies the free-field response of a pressure microphone is non-directional and equal to the pressure response. This contrasts with the behaviour at high frequencies where the microphone reflects the incident sound and the pressure is increased relative to its value in the unobstructed field. At normal incidence (0 degrees) the incident and reflected waves have approximately the same amplitude and the pressure on the diaphragm is twice that in the incident wave. The free-field sensitivity is then 6 dB greater than the pressure sensitivity. At grazing incidence (90 degrees) the presure magnification due to diffraction is negligible. At very high frequencies the phase of the sound pressure is not uniform over the diaphragm and accordingly the sensitivity is reduced. This effect adds to the directivity of the microphone but is usually unimportant because it generally occurs at frequencies outside the normal working range.

1.7.4 Calibration of pressure microphones

The pressure sensitivity of a microphone can be determined by applying a known pressure to the diaphragm. This is conveniently done using a *pistonphone*, whereby the microphone is coupled acoustically to a vibrating piston by means of a small cavity. The alternating pressure in the cavity is known from the amplitude of vibration of the piston and the volume of the cavity. Condenser microphones may also be calibrated by applying a known electrostatic force to the diaphragm by means of an external electrode which, together with the diaphragm forms a second parallel plate condenser. The instrument for doing this is known as an electrostatic actuator. The free-field sensitivity of a microphone can be found by comparison with a standard microphone or, absolutely, by the reciprocity method (Brüel and Kjaer 1967).

2 Anatomy and Physiology of the Ear

2.1 Introduction

The ear stands at the input to the auditory system; its function is to receive the acoustic stimulus and to initiate the corresponding activity in the auditory nervous system where analysis and subjective interpretation of the stimulus take place. Although part of the analytical process occurs within the ear itself, the role of the ear is primarily to convert the acoustic stimulus into suitably coded neural events and this is achieved by means of a series of acoustic and mechanical elements which in turn modify the incoming signal.

Although anatomy and physiology (structure and function) are intimately related, it is easier to understand the nature of a complex organ such as the ear by first considering its anatomy in isolation. Having obtained a clear mental picture of the overall structure, it is possible to examine the function of the component parts. The following description is a guide to the anatomy, highlighting those features which are of particular functional significance. For further details the reader is advised to consult the relevant anatomical texts provided in the bibliography at the end of this chapter.

2.2 Anatomy of the Human Ear

The ear consists of three anatomical units: the outer ear (sound receiver), the middle ear (mechanical transformer) and the innner ear (frequency analyser and transducer) (see figure 2.1). The outer and middle ear together form the so-called conductive system whereby the incident acoustic stimulus is transmitted to the inner ear. The inner ear is the interface with the nervous system; it transforms the acoustic stimulus into neural impulses which are transmitted in a complex system of interacting neural pathways linking the two ears and the brain. Most of the ear is contained within a part of the base of the skull

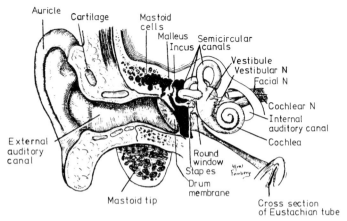

Figure 2.1 A semi-diagrammatic drawing of the ear. The inner ear is shown with the temporal bone cut away to reveal the semicircular canals, the vestibule and the cochlea. The cochlea has been turned slightly from its normal position to show the coils more clearly. The internal auditory canal is schematic. The middle ear muscles are not shown. From Davis and Silverman (1970) *Hearing and Deafness* © Holt, Rinehart and Winston.

known as the temporal bone. This is composed mainly of low-density material containing numerous air cells, but within its inner or petrous part is a region of extremely hard bone—the periotic capsule or bony labyrinth—containing the inner ear.

2.2.1 The conductive system

The external visible portion of the ear, the auricle or pinna, has a central bowl-shaped depression, the concha, which leads into the external auditory meatus. The meatus is a tube, circular or elliptical in cross section, some 5–7 mm in diameter and 25 mm in length, which runs in a slightly undulating course medially and horizontally. The auricle and outer third of the meatus are cartilaginous while the inner meatus is osseous, being part of the temporal bone. The cutaneous lining of the cartilaginous part contains glands which secrete cerumen (wax). The meatus is closed at its inner end by a conical membrane— the eardrum or tympanic membrane—which is inclined so that the floor and anterior wall of the meatus are longer than the roof and posterior wall.

Seen through the meatus, the drum appears to be circular except for the uppermost part which spans the notch of Rivinus. Thus the drum has two parts: the pars tensa which is the tympanic membrane proper and the pars flaccida or Shrapnel's membrane which is a thin, flaccid membrane in the notch just referred to. The drum is a composite of three layers having a total thickness of about 0·1 mm. The external layer is stratified squamous epithelium which is continuous with the lining of the meatus. The innermost layer is simple squamous epithelium, continuous with the mucosal lining of the ossicles and tympanic cavity. Between these layers is a relatively stiff lamina of connective tissue organised into radial and circular fibrils (Shimada and Lim 1971). This layer is absent in Shrapnel's membrane. In man, the pars tensa is not the stretched membrane that its name suggests, but is a rigid cone resembling the diaphragm of a loudspeaker. Movement of the cone is made possible by a flexible fold near its circumference opposite the axis of rotation of the malleus (figures 2.2 and 2.3), but elsewhere on the circumference movement is sufficiently small not to require a special elastic structure.

Behind the eardrum lies the tympanic or middle ear cavity—an irregular space having a volume of approximately 2 ml. The cavity has two parts: the tympanic cavity proper being that part opposite the eardrum, and the epitympanic recess which is situated above the drum. The epitympanum communicates posteriorly with the mastoid antrum and mastoid air cells. The lower part of the tympanic cavity is connected anteriorly through the Eustachian tube to the post-nasal space. The latter is at atmospheric pressure, and a valve mechanism within

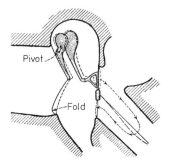

Figure 2.2 Schematic representation of the tympanic membrane, the ossicles and the basilar membrane. The stapes is drawn at right angles to its normal position to show its motion more clearly. From Davis and Silverman (1970) *Hearing and Deafness* © Holt, Rinehart and Winston.

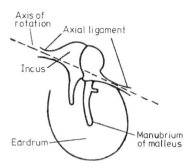

Figure 2.3 Lateral view of the eardrum and ossicles showing the axis of rotation. From Békésy *Experiments in Hearing*. Copyright © 1960. Used with permission of McGraw-Hill Book Company.

the Eustachian tube opens intermittently to cancel any static pressure difference that may develop across the eardrum. Most of the inner, or medial wall of the tympanic cavity is formed by the exposed outer surface of part of the bony labyrinth of the inner ear. This structure, which will be described presently, has two windows which open into the middle ear cavity.

The middle ear contains three small bones—the malleus, incus and stapes—which form a system of levers linking the drum to the inner ear. The largest of these bones, the malleus, has a process known as the handle or manubrium embedded in the tympanic membrane. The handle is slightly inclined to the vertical and lies radially in the drum, terminating at the centre (or umbo) of the cone. The malleus also has a neck and a head which lie above the drum in the epitympanic recess. The head bears an articulatory facet which forms a joint with a corresponding indentation in the body of the incus. This joint has little or no movement and mechanically the malleus and incus act as a single structure. From the body of the incus in the epitympanum a long process approximately parallel to the handle of the malleus runs downward into the tympanic cavity proper to make a movable joint with the head of the stapes. The latter is carried on a short neck which divides into two arms, the anterior and posterior crura. These terminate in an elliptical footplate lying in a plane approximately parallel to that of the eardrum, with the long axis of the ellipse horizontal. The footplate is secured by an elastic annular ligament in the oval window (fenestra vestibuli) which opens into the inner ear. The three ossicles are suspended within the tympanic cavity by a number of ligaments of which the most important are a pair running anteriorly from the short anterior process of the malleus and posteriorly from the short process of the incus. These attachments are in line and form the axis of rotation of the ossicular system. The axis is roughly horizontal and thus perpendicular to the manubrium and the long process of the incus, and it passes through the centre of mass of the complete ossicular system.

The middle ear has two muscles: the stapedius and tensor tympani, whose tendons attach to the neck of the stapes and the upper end of the manubrium of the malleus, respectively. The stapedius muscle is housed in a hollow pyramid of bone projecting from the posterior wall or the tympanic cavity. Contraction of the muscle pulls the head of the stapes posteriorly. The tensor tympani runs above the Eustachian tube parallel to the plane of the eardrum. Its tendon turns outwards through a right angle as it enters the tympanic cavity, so that the connection with the malleus is perpendicular to the drum. Thus the contraction of the tensor tympani draws the drum inwards.

2.2.2 The inner ear

The inner ear is a complex structure, both grossly and microscopically, which can best be understood if it is recognised as a bony container (the osseous labyrinth or periotic capsule) with soft tissue contents (the membranous labyrinth or otic capsule). The bony labyrinth is filled with a watery fluid called perilymph whose composition and electrolytic properties are similar to those of extracellular fluid. Suspended in the perilymph is the membranous labyrinth, which itself contains a fluid called endolymph. Endolymph, which is viscous, is electrolytically similar to intracellular fluid. The membranous labyrinth serves two functions: hearing and equilibration. The equilibratory end-organs are contained in three membranous semicircular canals and two membranous sacs: the utricle and saccule. The canals open into the utricle which is connected by the slender utriculo-saccular duct to the membranous cochlea, or cochlear duct, containing the organ of hearing. A description of the equilibratory or, as it is usually called, vestibular, system is outside the scope of this book. The reader seeking information on this subject is referred to works by Groen (1956) and Kornhuber (1974).

The part of the bony labyrinth containing the hearing apparatus is helical, resembling the shell of a snail from which its name, cochlea, is taken (figure 2.4). It lies in front of the vestibular part with its central axis, the modiolus, approximately horizontal and at 45 degrees to the

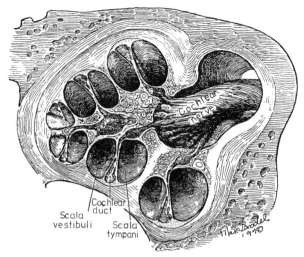

Scala
vestibuli
Cochlear
duct
Scala
tympani

Figure 2.4 The cochlea seen in a section through the modiolus.
From a drawing by Max Brödel 1940 *Year Book of the Eye, Ear, Nose
and Throat*. The original illustration is the collection of works by Max
Brödel, The Department of Art as Applied to Medicine, The John
Hopkins University School of Medicine.

external meatus. The helix has approximately $2\frac{1}{2}$ turns, and its basal
coil opens posteriorly into a bony chamber called the vestibule which
in turn leads to the osseous semicircular canals. The modiolus of the
cochlea is hollow and is principally occupied by the acoustic division of
the auditory nerve (VIII cranial nerve). It also forms the deepest part
of the internal auditory meatus which is a bony foramen in the petrous
part of the temporal bone, opening into the interior of the skull. Both
acoustic and vestibular divisions of the auditory nerve pass from the
brain to their respective end-organs in the inner ear via the internal
meatus, as does the blood supply to the labyrinth.

The bony cochlea is divided internally into two spiral chambers, the
scala vestibuli and scala tympani, by a partition which winds along the
entire length of the helix. At the apex a small opening in the partition
known as the helicotrema allows perilymph to pass between the two
compartments. At the basal end, the scala vestibuli opens into the
vestibule but the scala tympani terminates at the round window
(fenestra cochleae) which communicates with the tympanic cavity. The
round window is sealed by a flexible membrane (membrana secundum)
and is located at the upper end of a small recess in the medial wall of
the tympanic cavity. It is inclined downwards and backwards so as to

be almost perpendicular to the oval window. The oval window, which is filled by the stapedial footplate, opens from the tympanic cavity into the vestibule. Movement of the stapes displaces perilymph and the change in volume is transmitted to the scala vestibuli, where it is compensated by deformation of the cochlear partition and a yielding of the round window membrane.

The inner part of the cochlear partition is a thin bony shelf, the osseous spiral lamina, projecting from the modiolus. The partition is completed by a flexible membrane, the basilar membrane, which joins the outer wall of the cochlea in the spiral ligament. The osseous spiral is thickened on its vestibular surface to form the limbus from which Reissner's membrane runs diagonally to the outer wall. The triangle formed by Reissner's membrane and the tissues lining the spiral ligament and cochlear partition is the cross section of the membranous part of the cochlea. This forms the auditory division of the membranous labyrinth and is known as the cochlear duct or scala media. The duct ends blindly at the apex of the spiral while, at the basal end, it joins the saccule via the ductus reuniens. Like the rest of the membranous labyrinth, it contains endolymph.

Within the cochlear duct, supported on the partition and running the whole length of the spiral, is a ridge of neuro-epithelial tissue known as the organ of Corti. This is the end-organ of the auditory system; it is the final transducing mechanism for the detection of vibratory motion in the membranous part of the cochlear partition. The key feature of the organ of Corti is the presence of hair cells, so named because they carry stereocilia projecting from the ends of the cell bodies. These cells are innervated by the terminal fibres of the auditory nerve. Vibration of the cochlear partition communicated to the hair cells produces minute displacements of the cilia relative to the cell bodies and this action initiates the neural activity.

The organ of Corti is usually described as it appears in cross section, turned through 90 degrees so that the apex of the helix is uppermost (figure 2.5). It consists of the sensory cells and supporting structures. The central support is the tunnel of Corti—a spiral structure, triangular in cross section, having the basilar membrane as its base. The sides of the triangle are formed by rigid structures known as the pillars of Corti. At the vertex of the triangle the heads of the pillar cells bend outwards and broaden into flat horizontal plates. The sensory cells lie outside the tunnel disposed in a single inner row, and three (or occasionally four) outer rows. The inner cells, where they face the

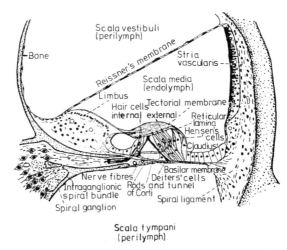

Figure 2.5 Cross section of the cochlear partition of a guinea pig in the lower part of the second basal turn. From Davis *et al* (1953). Reproduced with permission of The American Institute of Physics.

modiolus, are supported by cuboid epithelial border cells of an inner sulcus formed by the tympanic and vestibular lips of the spiral limbus. The outer hair cells rest on Deiters' cells. These cells send up phalangeal processes which pass through the space between the sensory cells (Nuel's space) and broaden into horizontal plates. Beyond Deiters' cells is a further supporting structure formed by the cells of Hensen and Claudius.

The uppermost part of the organ of Corti is a mantle known as the tectorial membrane. The membrane, which consists of fibrous material embedded in a gelatinous matrix, is attached to the vestibular surface of the limbus and stretches above the sensory epithelium to be supported externally by Hensen's cells. In surface view, with the tectorial membrane removed, a layer known as the reticular lamina becomes visible. This consists of the heads of the pillar cells and the phalangeal plates of Deiters' cells. Embedded in the lamina are the vestibular ends of the outer hair cells. In the same plane as the reticular lamina are the vestibular faces of the inner hair cells and the supporting border cells of the inner sulcus.

Each hair cell projects three rows of stereocilia upwards towards the tectorial membrane. From above, the cilia of the inner hair cells appear as a line running along the spiral but on each of the outer cells they lie in a U- or V-formation with the vertex of the V directed outwards. On each cell, the cilia of successive rows are of different lengths so that each group of hairs resembles a leaf spring. The outermost hairs are the longest and only these reach the tectorial membrane to which they are firmly anchored. It is thought that this arrangement gives greater mechanical stiffness for inward, as compared with outward, bending of the cilia.

The sensory cells are innervated by afferent and efferent fibres of the cochlear nerve. The afferent fibres have their cell bodies in the spiral ganglion within the modiolus and pass through the osseous spiral lamina to the organ of Corti. Connections to the external hair cells pass through the tunnel of Corti which, unlike the rest of the scala media, contains perilymph or perhaps some other fluid of similar electrolytic composition. Afferent fibres of three types have been identified: specific radial fibres, each of which contacts only a very few neighbouring inner hair cells; spiral fibres which each innervate many external hair cells in a long segment of the organ of Corti; and fibres which branch at irregular intervals to innervate several widely separated inner hair cells (Lorente de Nó 1976). Further information about the auditory nervous system is given in Chapter 4.

When looking at anatomical drawings, it should be remembered that the ear, excluding the external part, is built on a very small scale. Man is by no means the smallest of mammals, yet his ear, if it could be removed, would stand in an area the size of a thumbnail. A table giving the dimensions of parts of the ear is to be found in Wever and Lawrence (1954).

2.3 Physiology of the Conductive System

For monaural hearing, the ear is a pressure-sensitive detector analogous to a pressure microphone, but binaurally the auditory system is not a simple pressure receiver because the distance between the two ears is a significant fraction of the wavelength at all audible frequencies. The binaural properties of the auditory system will be considered in §3.8.

The head may be treated as a rigid spherical obstacle which disturbes the sound field and raises the sound pressure at the entrance to the meatus. The problem of diffraction by a sphere has been solved theoretically by Ballantine (1928) and the result for a sphere of radius

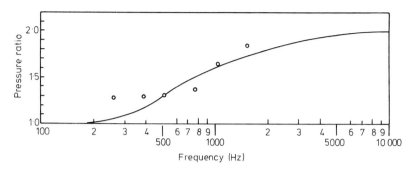

Figure 2.6 Pressure at the entrance to the meatus relative to the pressure in a previously undisturbed field due to diffraction by the head. The sound waves are incident on the side of the head. The line shows Ballantine's calculations; the points are experimental values obtained by Langenbeck (1931). From Békésy *Experiments in Hearing.* Copyright © 1960. Used with permission of McGraw-Hill Book Company.

90 mm representing the head is shown in figure 2.6. In this analysis it is assumed that the disturbance of the sound field by the auricle is negligible. Békésy† has shown that this assumption is justified, at least for frequencies up to 3 kHz, but diffraction by the auricle probably assists the location of high-frequency sounds.

The external meatus may be treated as an acoustic transmission line‡ having distributed inertance and compliance. The line is 'loss-less' because resistance due to viscosity of the air and losses by conduction radially through the meatal wall are negligible.

From equations (1.24) and (1.25) it is evident that the inertance and compliance per unit length of the meatus are $M = \rho_0/S$ and $C = S/\rho_0 c^2$, respectively, where S is the cross sectional area of the meatus. The characteristic impedance of the line is a resistive impedance given by

$$Z_0 = \sqrt{M/C} = \rho_0 c/S. \qquad (2.1)$$

† Georg von Bekésy's contributions to the theory of hearing are so prolific that references to specific publications of his will not be given. A detailed account of his work up to 1960 is presented in his book *Experiments in Hearing*, which includes English translations by E G Wever of many articles which were originally published in German. The author's bibliography lists 83 works published between 1928 and 1958. Bekésy was awarded the Nobel prize for Medicine and Physiology in 1961.

‡ The theory of transmission lines is usually found in texts on electrical engineering, such as King *et al* (1945) and Seshadri (1971).

For a typical meatus with an area of $44 \, \text{mm}^2$ the value of Z_0 is $9 \cdot 2 \times 10^6 \, \text{N s m}^{-5}$ (Zwislocki 1975). The transmission coefficient at the drum is (from equation (1.34)),

$$\alpha_T = \frac{4 \, |Z_0 Z_D|}{|Z_0 + Z_D|^2},$$ (2.2)

and reflections occur unless the terminating impedance, Z_D, at the drum is equal to Z_0. It is difficult to make precise measurements of Z_D at all frequencies, but it appears that the termination is never quite ideal, with the result that the pressure at the drum is greater than the pressure at the entrance to the meatus. The gain is a maximum at the principal resonance of the meatus which occurs at about $2 \cdot 5 \, \text{kHz}$. The pressure magnification due to the combined effects of the head and the outer ear is shown in figure 2.7.

The specific acoustic impedance of the perilymph at the oval window is about 300 times greater than that of the air in the meatus, and this disparity is resolved by the transformer action of the middle ear. A second function of the middle ear is to direct the incoming energy preferentially to the oval window and thus create a pressure *difference* between this and the round window. The middle ear transformer can be represented schematically by two pistons (eardrum and stapes) connected by a lever (malleus and incus). Although this representation is an over-simplification, it is interesting to see the degree of impedance matching that can be achieved. Let the acoustic impedances at the

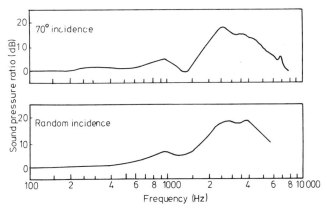

Figure 2.7 Pressure at the eardrum relative to pressure in a previously undisturbed free field. From Pollack (1949) *Am. J. Psychol.* **62** 412–7 (The University of Illinois Press).

eardrum and stapes be Z_1 and Z_2, respectively. Then if p_1 and U_1 are the pressure and the volume velocity at the eardrum, and p_2 and U_2 are the corresponding quantities at the stapes, by definition

$$Z_1 = p_1/U_1 \quad \text{and} \quad Z_2 = p_2/U_2.$$

According to the reciprocity theorem the transfer impedances are equal, that is, $p_1/U_2 = p_2/U_1$. Therefore

$$Z_1/Z_2 = (p_1/p_2)^2 = (U_2/U_1)^2 . \tag{2.3}$$

For a simple transformer, U_1 and U_2 are in the same phase, and impedance matching is only possible if the vectors representing the complex impedances Z_1 and Z_2 are concurrent. In the ear the ideal termination of the meatus would be resistive as already discussed. Zwislocki (1965) has shown that the impedance at the oval window is also resistive, and thus in principle perfect matching is possible. The vector notation in equation (2.3) can therefore be omitted and the terms then represent the magnitudes of the pressure or volume transformations required for impedance matching.

In order to estimate the volume transformation in the middle ear it is necessary to know the relative areas of the eardrum and stapes and the mechanical advantage afforded by the ossicular lever. At frequencies up to 2·4 kHz the central zone of the drum moves as a rigid diaphragm, stiffly connected to the manubrium. Békésy found that the area of this part was effectively 18·7 times that of the stapedial footplate. In the ossicular lever the incudo-malleolar joint is non-articulatory, but movement does occur in the incudo-stapedial joint. The resulting motion of the stapes is the combination of a lateral piston-like movement and a rotation about a vertical axis near the posterior border of the footplate.

The rotary component is due to non-uniformity of the annular ligament which is stiffer posteriorly than anteriorly† (Kirikae 1973). The mechanical advantage of the ossicular lever is only about 1·3. This, multiplied by the ratio of the areas of drum and footplate, gives a volume transformation of 24 and hence an impedance transformation

---

† According to Békésy the motion of the stapes at normal intensities is a rocking motion about a posterior axis—a piston-like, lateral component is not mentioned. Bel *et al* (1976) refute this traditional view. They claim that the motion is a rocking about an anterior axis and that this is due, not to non-uniformity of the annular ligament, but to the shape of the border of the footplate and the rim of the oval window which they describe as helicoidal.

of 576. According to the best available data, the acoustic impedance at the oval window is $3\cdot5 \times 10^{10}\,\mathrm{N\,s\,m^{-5}}$ which, divided by 576, gives a transformed impedance at the drum of $6\cdot1 \times 10^{7}\,\mathrm{N\,s\,m^{-5}}$. The mismatch between this and the ideal termination, $9\cdot2 \times 10^{6}\,\mathrm{N\,s\,m^{-5}}$ (equation (2.1)), yields a transmission coefficient of $0\cdot46$ ($-3\cdot4\,\mathrm{dB}$) for the transfer of energy from the meatus to the cochlea.

In reality, the middle ear is not the simple transformer described above because its components have significant mass and stiffness. Figure 2.8 shows the measured acoustic impedance at the eardrum in normal human subjects. Below about 800 Hz the impedance is dominated by a negative reactance due largely to the stiffness of the drum, the ossicular system and the cushion of air trapped in the tympanic cavity. At higher frequencies this reactance is negligible except above 6 kHz, where the mass of the ossicles contributes a small positive reactance. The resistive part of the impedance is approximately constant, having a value of about $3 \times 10^{7}\,\mathrm{N\,s\,m^{-5}}$, or roughly half the transformed impedance of the cochlea.

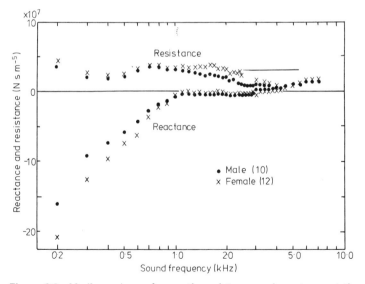

Figure 2.8 Median values of acoustic resistance and reactance at the human eardrum for males and females. Above 3 kHz direct measurement of resistance is unreliable—an estimated value is shown by the horizontal line. From Zwislocki 1975 *The Nervous System* (New York: Raven).

A method of analysing the performance of the middle ear is to represent it in terms of its equivalent circuit. The circuit is a network of acoustic or mechanical elements, or their electrical equivalents, with an input representing the meatus and an output representing the cochlea. The problem is to determine the transfer function (ratio of output to input power) of the network and to see how this is related to the behaviour of the various components. It is obviously difficult to know what numerical values should be assigned to the elements of the network, but some information can be obtained indirectly from measurements made on ears in which parts of the middle ear mechanism are modified or removed as a result of disease or surgical intervention. This analytical procedure has been used by several workers, notably Møller (1960) and Zwislocki (for a review see Zwislocki 1975). The results suggest that the frequency dependence of auditory thresholds† is determined principally by the transfer function of the conductive system (figure 2.9).

The transformer action of the middle ear could be accomplished simply by a single bone coupling the oval window and the drum, and in fact such a mechanism, known as a columella, is found in amphibians, reptiles and birds. According to Békésy, the need for the complicated

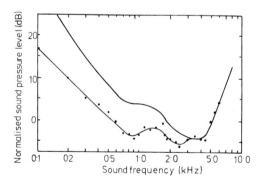

Figure 2.9 Comparison between the overall outer and middle ear transfer functions (circles and lower curve) and the median threshold of audibility. The relative positions of the two curves on the vertical scale are arbitrary. From Zwislocki 1975 *The Nervous System* (New York: Raven).

---

† Auditory thresholds are discussed in Chapter 3. In the context of figure 2.9 the threshold of audibility is the intensity of the minimum audible free field.

middle ear arrangement in mammals becomes apparent when bone conduction is considered. This mode of conduction occurs when the inner ear is stimulated directly by vibration transmitted through the skull itself. Normally bone conduction is unimportant because the skull has a relatively high mass and rigidity so that its vibratory response in a sound field is small. The transmission loss for airborne sound heard directly through the skull is at least 60 dB. However, the act of speaking (phonation) produces significant vibration of the head, particularly in the vertical plane. The ossicular system, since it is contained within the temporal bone, must vibrate with the head and must therefore be involved to some degree in bone conduction. A feature of the middle ear mechanism in man is that this involvement is small. The principles of bone conduction are discussed by Naunton (1963) and Tonndorf (1970).

2.4 The Middle Ear Muscles

The middle ear muscles and their tendons form part of the mechanism for supporting and stabilising the ossicular chain. Unless stimulated, the muscles exert no force other than their normal resting tonus. Contraction of either or both of the muscles stiffens the ossicular system and thus increases the acoustic reactance at the eardrum; changes in the resistive component of the acoustic impedance are small and variable (Dallos 1964). The middle ear response occurs as an involuntary reflex. The stapedius muscle, which is innervated by a branch of the VII cranial nerve (facial nerve), is active when either ear is exposed to a loud sound. The tensor tympani is innervated by a branch of the V cranial nerve (trigeminal nerve). Its response is part of a general 'startle' reaction. Contraction of the tensor tympani may be elicited by a number of stimuli, including tactile and electrical stimuli, but it probably does not respond to pure acoustic stimulation unless the subject is also startled by the sound. Methods of obtaining middle ear reflexes will be described in Chapter 6. The stapedius response is more rapid than that of the tensor tympani. Thus Casselbrant *et al* (1977) found a latency[†] of about 100 ms for the stapedius muscle, compared with 650 ms for the tensor tympani.

[†] Latency is the interval from the presentation of the stimulus to the appearance of the response. It depends on the intensity of the stimulus and the method of observing the response. For a detailed account of the dynamics of the tympanic muscles see Borg (1976).

Sustained contraction of the tensor tympani does not occur naturally; the tensor tympani response is a transitory inward displacement of the malleus and a concomitant tensing of the tympanic membrane. The action of the stapedius muscle is more subtle; its contraction, which is sustained for the duration of the stimulus, pulls the head of the stapes posteriorly. This probably produces a translational or 'gliding' displacement in the incudo-stapedial joint which, according to Békésy, has greatest freedom of movement in the direction of the axis of the footplate. At the tympanic membrane the net effect of this activity is an increase in acoustic reactance, together with small displacement of the drum which may be either inwards or outwards according to subject (Casselbrant *et al* 1977).

The functional importance of the middle ear musculature is uncertain. The freedom of the ossicular system must necessarily be restricted in order to prevent the formation of sub-harmonics (Békésy) and in this context the stabilising function of the muscle attachments may be significant. Békésy considered that the most important function of the tympanic muscles was to assist the ligaments in preventing the separation of the inter-ossicular joints during the transmission of high-intensity and high-frequency sounds. The acoustic (stapedius) reflex provides a limited protection of the inner ear against sustained loud noise, but the protection is confined to low frequencies which are in fact the least harmful. For a brief review of theories of the functional significance of the middle ear muscles see Jepsen (1963).

2.5 Physiology of the Inner Ear

The working of the inner ear can best be understood if the cochlea is represented schematically as shown in figure 2.10. In this diagram the spiral form is ignored and the bony coil is shown as a straight tube. The membranous duct is reduced to a single lamina (the basilar membrane) which divides the tube into the two scalae. The function of the inner ear is then considered in terms of the transverse displacement of the membrane which proceeds as a travelling wave running from base to apex. The amplitude and velocity of the wave are controlled by the properties of the membrane.

Our knowledge of the inner ear mechanism is based largely on the brilliant experimental work of von Békésy. His research began over 50 years ago with the construction of mechanical models of the cochlea which were dimensionally scaled according to physical parameters

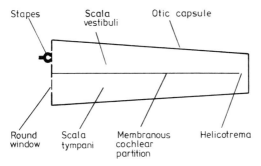

Figure 2.10   Schematic representation of the inner ear.

measured on temporal bone preparations from cadavers. Using the models he was able to study the simulated behaviour of the cochlear partition. Subsequently (in the early 1940s) Békésy succeeded in making direct observations of the human cochlea. To avoid damage to the delicate tissues, it was necessary to open the bony capsule under water. The membranous partition was then observed using a microscope fitted with a water-immersion objective. The results showed that the mechanical models had provided a correct description of the real ear. In summary, Békésy's findings were as follows.

(1) The width of the basilar membrane increases from about 0·08 mm at the base to 0·5 mm at the apex. The acoustic compliance per unit length of the membrane increases exponentially from base to apex (base 10^{-12} apex 10^{-10} m^4 N^{-1}). There is no measurable tension in the membrane.

(2) The pattern of vibration in the models was not critically dependent on the viscosity of the fluid or the depth of the channels representing the cochlear chambers.

(3) The travelling wave always runs with decreasing velocity from base to apex, irrespective of the point of stimulation (normally the oval window).

(4) As the wave progresses, its amplitude increases at first but after passing through a maximum it declines rapidly (figure 2.11).

(5) The location of the maximum moves systematically from apex to base as the frequency of stimulation is increased (figure 2.12).

(6) As the wave proceeds an increasing phase lag accumulates in the displacement of the basilar membrane relative to the displacement of the stapes (figure 2.12).

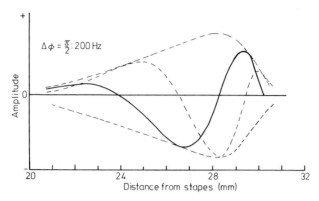

Figure 2.11 Vibration of the human cochlear partition at 200 Hz. The travelling wave is shown at two successive instants corresponding to a phase separation of $\pi/2$. The outer broken line is the envelope of the waveform. From Békésy *Experiments in Hearing*. Copyright © 1960. Used with permission of McGraw-Hill Book Company.

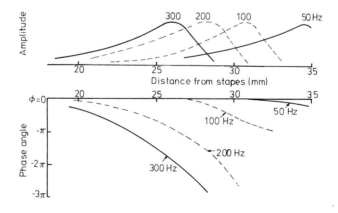

Figure 2.12 Amplitude and phase of vibration of the cochlear partition. The phase is that of the displacement of the partition referred to the displacement of the stapes. The maximum amplitude occurs at a point where the phase change is approximately -2π. From Békésy *Experiments in Hearing*. Copyright © 1960. Used with permission of McGraw-Hill Book Company.

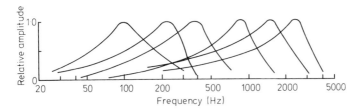

Figure 2.13 Amplitude of vibration at six points on the cochlear partition as a function of frequency. From Békésy *Experiments in Hearing*. Copyright © 1960. Used with permission of McGraw-Hill Book Company.

(7) The vibration of the partition is heavily, but not critically, damped.

From (5) it follows that a set of curves can be constructed showing the frequency response at specified points on the partition (figure 2.13). These are sometimes called 'resonance' curves, although the phenomenon is not resonance. Thus the cochlear mechanism acts as a mechanical filter through which each region of the basilar membrane is selected for maximum stimulation at a particular frequency. In terms of the transverse displacement of the membrane, the tuning is not very sharp and seems insufficient to explain the discrimination of the auditory system which can recognise frequency changes of 0·3%. Békésy believed that the selectivity of the mechanical filter in the cochlea was enhanced by neural processing of the signals coming from the hair cells. This form of processing, which is a general feature of the nervous system, can be demonstrated for visual and tactile sensations. For example, if a non-uniform tactile stimulus is distributed over an area of skin on the forearm, the pattern of the sensation has a higher contrast than that of the stimulus (Békésy 1970). The sensation is strengthened in regions of high stimulation and inhibited elsewhere. This sort of processing applied to the cochlea would reduce the effective bandwidth of the tuning mechanism. Békésy's hypothesis is indirectly supported by the results of electrophysiological measurements in which the activity of individual units of the auditory nerve is examined by the insertion of microelectrodes (Kiang 1975). It appears that the frequency response of single nerve fibres is highly selective (figure 2.14). Opponents of the hypothesis of neural processing point out that so far the appropriate neural mechanism has not been identified (Kim and

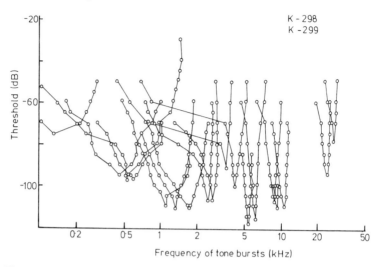

Figure 2.14 Tuning curves for single fibres of the cochlear nerve in anaesthetised cats. Data from two cats are shown. The ordinate is the stimulus intensity required to produce a detectable change in the activity of the neurone. From Kiang 1975 *The Nervous System* (New York: Raven).

Molnar 1975), although it is perhaps doubtful whether such a mechanism could be recognised even if it were uncovered. A further consideration is that stimulation of the hair cells is not simply a matter of lateral displacement of the basilar membrane, but probably involves a radial shear between the basilar and tectorial membranes. This occurs because the basilar membrane is firmly attached to the bony cochlea along both its margins, whereas the tectorial membrane is secured unilaterally by its attachment to the limbus. The tuning curves for shear displacement are sharper than those shown in figure 2.13 (see Tonndorf 1970).

2.6 Physical Theory of the Cochlear Mechanism

Following Ranke (1950), it has been shown by Siebert (1974) that the travelling wave pattern can be accurately reproduced in a mathematical model. The physical basis of the model is extremely simple but the ensuing mathematics is rather difficult. In the model the cochlea is represented by a rectangular fluid-filled box, divided into two equal channels of depth d by a flexible partition. The oval and round windows are replaced by equivalent sources at one end of the box, while at the opposite end, a second pair of sources represents reflection at the apex of the cochlea. The mechanical properties of the partition itself are expressed in terms of its admittance which is a function of the distance x from the 'stapes' and depends on the stiffness, mass and resistance of the membrane. In Siebert's analysis the terms of the admittance function were derived empirically by selecting these parameters to make the simulated wave pattern on the model agree as closely as possible with Békésy's experimental observations. Siebert noted that while there were no independent measurements of mass and resistance, the optimum stiffness function was in close agreement with Békésy's findings. The fluid in the model is assumed to have no viscosity, so that the only damping is that provided by the resistive component of the membrane admittance.

In order to obtain a workable mathematical solution it is necessary to make one of two approximations. Thus it is assumed that at all frequencies of interest the wavelength is either large or small compared with $2\pi d$, at all points on the partition. The latter is the so-called short-wave approximation and it results in a wave motion that is not affected by the boundaries of the container. Siebert pursued the short-wave analysis and showed that the wave form and frequency response of the basilar membrane can be simulated with encouraging accuracy. The principal failure of the model is that it does not correctly reproduce the observed phase change. Unlike the long-wave models, the short-wave approximation gives rise to a highly dispersive system.

Long-wave models have been developed by numerous investigators, notably Zwislocki (1950). In Zwislocki's analysis, which is not difficult to follow, the velocity of the travelling wave is given by $V = \sqrt{q/\rho C}$, where ρ is the density of the perilymph, C is the acoustic compliance per unit length of the partition, and q is a constant relating to the cross sectional areas of the two cochlear chambers. Zwislocki (1965, 1975) has used this model to estimate the impedance at the oval window from the relationship

$$Z^2 \simeq \frac{S_t S_v}{C(S_t + S_v)},$$

where S_t and S_v are the cross sectional area of the scala tympani and scala vestibuli. The result is in good agreement with Békésy's direct measurement of impedance corrected for post-mortem changes.

Bibliography

The following works contain detailed descriptions of the anatomy of the ear.

Anson B J and Donaldson J A 1967 *The Surgical Anatomy of the Temporal Bone and Ear* (Philadelphia: W B Saunders)

Bischoff A (ed) 1970 *Ultrastructure of the Peripheral Nervous System and Sense Organs* (Stuttgart and London: Georg Thieme Verlag and Churchill)

Maximow A A and Bloom W 1942 *A Textbook of Histology* 4th edn (Philadelphia: W B Saunders)

Romanes G J (ed) 1964 *Cunningham's Textbook of Anatomy* (London: Oxford University Press)

Soudjin E R 1976 Scanning Electron Microscopic Study of the Organ of Corti *Ann. Oto-Rhino-Laryngol.* **85** suppl. 29

Warwick R and Williams P L (eds) *Gray's Anatomy* 35th edn (London: Longman)

Wolff D, Bellucci R and Eggston A 1957 *Microscopic Anatomy of the Temporal Bone* (Baltimore: Williams and Wilkins)

3 The Nature of Hearing

3.1 Introduction

The performance of the auditory system in its entirety may be studied by examining the relationship between the physical attributes of a sound and the corresponding auditory sensation. Of particular interest is the ability of the ear to recognise the presence of very weak sounds. The intensity of the least audible sound is known as the threshold of hearing. The existence of a measurable quantity that can be called a threshold is of fundamental importance in almost every branch of auditory theory and is the *sine qua non* of diagnostic audiometry. This aspect of auditory performance will therefore be considered first.

3.2 Auditory Thresholds

In order to investigate the limits of his hearing, a person may be presented with a number of sounds (often pure tones) of varying intensity and asked to respond in some way to those sounds which he can hear. In this manner the minimum intensity required to produce the sensation of hearing can be determined. The probability that a given auditory stimulus will produce a response depends primarily on the magnitude of the stimulus. It also depends on the disposition of the subject, namely, his alertness, his willingness and ability to listen, his motivation, and so on. According to the threshold theory the probability of obtaining a genuine response to a subliminal stimulus is necessarily zero, for such a stimulus is by definition inaudible. Any such response is therefore regarded as an error on the part of the observer. Above the threshold the probability of obtaining a response becomes finite and rises rapidly with increasing intensity of the stimulus. Thus the probability of hearing a weak sound is approximately a unit step function at threshold intensity. The fundamental weakness of the threshold hypothesis is that it is never possible for a listener to experience complete silence because the auditory system, like any other receiver, suffers from inherent noise. This might be called physiological noise. It is due to such things as Brownian motion at the eardrum and in the cochlear fluids (de Vries 1952), spontaneous electrical activity in the nervous system, and vascular and respiratory noise. The ability to detect a weak acoustic signal therefore depends on those attributes of the signal which distinguish it from the noise rather than on any threshold or quantum mechanism within the auditory system. The detection of weak signals or signals in the presence of added noise has been the subject of numerous psychophysical investigations in which an observer is required to locate the signal in one of a number of presentations of signal plus noise or noise alone (the noise need not be added deliberately—it may be physiological or inherent in the experimental apparatus). These experiments have clearly falsified the threshold hypothesis. They show instead that the observer's performance can be described in terms of a decision-making process in which the nature of the signal, the observer's experience (learning) and his criteria for affirming the presence of the signal are all contributory factors. A useful introduction to the theory of signal detectability and related psychoacoustic techniques is provided by Tanner and Sorkin (1972) and Clarke and Bilger (1973).

Despite its shortcomings, the threshold theory remains a valuable tool both in research and in clinical audiology. One reason for this is that the dynamic range of the ear is large compared with the intensity range in which threshold phenomena occur and, although different methods of estimating the threshold give different results, these differences are not usually significant. This is exemplified in figure 3.1, which shows thresholds for an interrupted tone obtained by the presentation of successively increasing or decreasing stimuli. The cumulative fraction of affirmative responses is shown in each case. It can be seen that the separation of the ascending and descending curves is less than 5 dB, and that for each method the probability of obtaining a positive response rises from zero to unity in a range of less than 10 dB. Of course grossly different methods of obtaining a threshold may give significantly different results. The usual procedure is to present the stimulus several times at the same intensity, each presentation lasting for about two seconds followed by a silent interval. The threshold may then be defined as the intensity for which a response is obtained to approximately half the number of presentations. Alternatively, the threshold may be approached from above and below and the two lowest intensities which give consistent responses averaged.

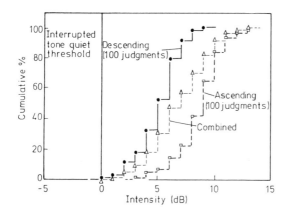

Figure 3.1 Thresholds for an interrupted 4 kHz tone presented at successively increasing and decreasing intensities. The ordinate shows the cumulative percentage of affirmative responses obtained in each case. From Hirsh *The Measurement of Hearing.* Copyright © 1952. Used with permission of McGraw-Hill Book Company.

Hearing thresholds can be measured for either free-field or earphone listening. The former gives the 'minimum audible field,' defined as the intensity of a previously undisturbed free field which is just audible when the listener is present. In this situation there is a pressure gain due to diffraction by the head and the normal transmission properties of the outer ear. For earphone measurements thresholds are expressed as the 'minimum audible pressure.' This is the sound pressure under the earphone measured in the entrance to the meatus. The transmission of sound in the ear canal is modified by the presence of an earphone and diffraction by the ear, of course, does not occur. Normal free-field thresholds are shown in figure 3.2. These values are for binaural listening with the subject facing a source of free progressive plane waves. Thresholds for monaural listening are about 3 dB higher than those for binaural listening (Shaw *et al* 1947), presumably because the signal-to-noise ratio in the auditory system is halved when one ear is occluded. Because of the importance of earphone measurements in audiometry there have been numerous attempts to establish a standard minimum audible sound pressure for 'normal' hearing (see Chapter 5). Pressure thresholds obtained by Dadson and King (1952) and Wheeler and Dickson (1952) were in good agreement with the results of earlier work (Sivian and White 1933) and formed the basis of BS 2497 (1954) (see figure 3.3). Over a large part of the audible range the frequency dependence of the free-field threshold is probably determined by the transfer function of the conductive system as described in Chapter 2. Since diffraction effects are absent when earphones are worn, the pressure threshold is generally higher than the free-field threshold and irregularities due to meatal resonance are less prominent. However, Munsen and Wiener (1952) could find no adequate explanation for the difference between the two thresholds at low frequencies. This difference became known as the 'missing 6 dB' and an explanation for it was provided nearly 20 years later by Anderson and Whittle (1971). They showed that the wearing of an earphone increased the physiological noise in the meatus. The noise, which is vascular in origin, has a small masking effect at low frequencies.

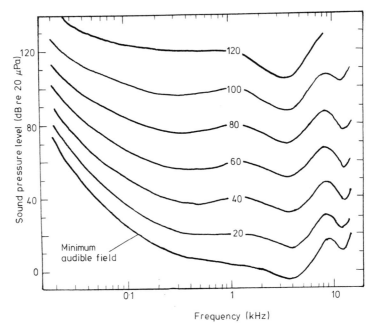

Figure 3.2 Minimum audible field and equal-loudness curves for binaural listening. Numbers on the curves denote loudness levels in phons. From BS 3383 (1961), reproduced by permission of BSI, 2 Park Street, London WIA 2BS, from whom complete copies can be obtained.

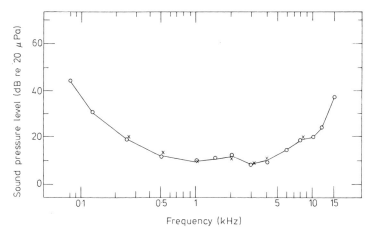

Figure 3.3 Minimum audible pressure at the entrance to the external auditory meatus for sound delivered monaurally through an earphone. Circles, Dadson and King (1952); crosses, Wheeler and Dickson (1952). Data taken from BS 2497 (1954), reproduced by permission of BSI. This standard has now been withdrawn and superseded by three parts issued in 1968, 1969 and 1972.

3.3 Loudness

Loudness is a subjective quality that depends on the intensity and frequency of a sound. The intensity relations among equally loud sounds of different frequencies can be found by a series of loudness balances with a reference sound, and in this way equal-loudness contours can be drawn on a graph of intensity against frequency (figure 3.2). Each of these curves may be given a numerical rating known as its loudness level on the phon scale. On this scale the loudness level of a sound is equal to the sound pressure level above $20\,\mu$Pa of a reference 1 kHz tone which is judged to have the same loudness as the sound in question. The zero-loudness curve (0 phons) is approximately equal to the minimum audible field. It lies in fact a little below the threshold shown in the diagram because a reference sound pressure of $20\,\mu$Pa has been used to define the phon, whereas the level of the free-field threshold (BS 3383) is $4\cdot2$ dB at 1 kHz. With increasing loudness there is a progressive change in the shape of the equal-loudness contours. For very loud sounds these curves are distinctly flatter than the threshold curve and at the limit of tolerable loudness frequency becomes unimportant. Sounds become uncomfortable at

120 phons and painful at 140 phons. Not surprisingly, these high intensities are injurious to the hearing mechanism.

The measurement of a physical quantity often involves a subjective assessment at some stage. For example, a photometric measurement may require the observation that one source of illumination has the same brightness as another. If the two sources have the same colour the measurement can be regarded as purely physical since the judgment of equal illumination could in principle be made by an artificial observer (a photocell). On the other hand, the judgment that two dissimilar sounds are of equal loudness is entirely subjective since it could not be made by any machine or scientific instrument. Nevertheless, accurately reproducible measurements of loudness level can be made—the standard error in the average judgment of ten observers is likely to be in the range 1–3 phons. The phon is not an absolute measure of subjective loudness but rather a means of comparing the loudness of different sounds. It is, however, possible to construct a scale of loudness which is completely subjective (Robinson 1957, Stevens 1955). The scale is based on judgments that one sound is half or twice as loud as another, or that it is half-way in loudness between one sound and another. There is considerable variability with such assessments, perhaps because the human observer is unaccustomed to making them and, furthermore, has no objective standard against which to measure his performance (he would have relatively little difficulty in mentally doubling or halving the distance between two points). The unit of loudness on the subjective scale is the sone. Doubling the loudness of a sound is equivalent to increasing its loudness level by 10 phons (BS 3383), The loudness in sones, S, is related to the level in phons, P, by

$$10 \log S = (P - 40) \log 2 . \tag{3.1}$$

For a pure tone at 1 kHz, $P = 10 \log I/I_0$, where I is the intensity of the tone at loudness level P, and I_0 is the threshold which, for the sake of this discussion, is assumed to coincide with the zero on the phon scale. Equation (3.1) can be written in the form

$$S = k(I/I_0)^n \tag{3.2}$$

where $k = 1/16$ and $n = \log 2 = 0\cdot3$. It is alleged that equation (3.2) is a general expression of a power law relating the subjective and physical magnitudes of a stimulus. The average value of the exponent depends on the nature of the sensation; for loudness it is 0·30, for brightness 0·33, for temperature 1·0, and so on (Stevens 1957, 1961).

The loudness scale, unlike the corresponding scales for other sensations, has a practical application. This is in the measurement of noise where it enables loudness to be estimated from objective recordings of sound pressure. The appropriate methods are described in BS 4198.

3.4 Pitch

Pitch is the sensory concomitant of frequency, and it has two distinct attributes. One is the sense of a continuous rise in 'tone-extent' with increasing frequency; and the other is recognised as recurrent musical intervals corresponding to specific frequency ratios. The latter is of special importance to musicians, and for those interested in music the article by Ward (1970) is particularly recommended.

The lowest frequency at which sound has a tonal quality is about 15 Hz. The sensitivity of the ear for low tones declines rapidly with decreasing frequency and the threshold of hearing occurs at increasingly high acoustic intensities. Significant distortion is present within the ear at high levels of stimulation and, if the frequency is sufficiently low, harmonics of the stimulus may be audible even though the fundamental is not. Raising the intensity of the stimulus results in a large increase in distortion but relatively little increase in the strength of the fundamental. The growth of each overtone is a power function of intensity with an exponent equal to the order of the harmonic (Wever *et al* 1940). According to Wever (1949) the subjective consequence of this disproportionate increase in the strength of the harmonics is reinforced by the changing sensitivity function of the ear which favours the reception of the higher-frequency components. While this may be true, it should be remembered that the threshold curve shown in figure 3.2 applies only to parameters of the external stimulus and not to harmonics generated within the ear itself. Below about 15 Hz sounds remain inaudible at their fundamental frequency whatever their intensity, and can only be sensed through the distortion products. Audibility extends downwards to at least 5 and possibly to 1 Hz. These very low frequency sounds are heard as a complex noise resembling 'the chugging of a reciprocating pump heard a long way off' (Wever and Bray 1937). Below 15 Hz the pitch of this noise rises with decreasing frequency in keeping with the above observations. At the other extreme, the limit of audibility is about 20 kHz, above which there is no sensation whatever.

Pitch depends to some extent on intensity. High-frequency tones become a little sharper as the intensity is raised, while the opposite occurs at low frequencies (Stevens 1935). For a discussion of this phenomenon see Wever (1949).

3.4.1 Periodicity pitch

The pitch of a pure sinusoid corresponds directly to its frequency, but the pitch of a complex sound is related to the periodicity in the envelope of the waveform. A famous demonstration of this phenomenon was provided by Seebeck (1841) using a siren to generate sounds consisting of a series of impulses, as shown in figure 3.4. For the waveforms labelled (a) and (b) the impulses are equally spaced with (b) having twice the repetition rate of (a) and sounding an octave higher. In (c) and (d) these patterns are combined, but (a) is given a displacement relative to (b) so that alternate pulses occur at unequal intervals. In (c) the displacement is small; in (d) it is large. When either (c) or (d) is sounded the two pitches previously associated with (a) and (b) are heard simultaneously. It is particularly important to note that the frequency spectra of (c) and (d) contain very little energy at the fundamental frequency of the lower tone, yet the corresponding pitch is clearly perceived.

Using a mechanical model, Békésy (1955, 1961) simulated the response of the cochlea to periodic stimuli. In the model the artificial basilar membrane was arranged to produce a tactile stimulus distributed along an observer's arm placed in contact with it. When the

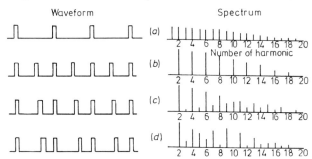

Figure 3.4 Seebeck's experiment. For explanation see text. After Schouten 1940 *Proc. K. Ned. Acad. Wet.* **43** 356–65 (Amsterdam: North-Holland).

model was driven sinusoidally the sensation was localised at the maximum displacement of the membrane and moved progressively from elbow to wrist as the frequency of the stimulus was increased. Thus pitch was represented by the location of the sensation. When the model was driven by a series of impulses, maxima in the membrane displacement occurred at a place corresponding to the periodicity of the input waveform as well as that corresponding to its principal Fourier component, and were perceived in a manner analogous to the auditory sensation. This suggests that in some circumstances the phenomenon of periodicity pitch has a physical explanation in terms of the place principle of pitch perception. It has also been suggested that periodicity pitch is due to distortion within the ear and that it has a physical basis in the existence of intermodulation products derived from components of the original stimulus. There is, however, ample evidence to show that this is not the case and that the presence of energy at the fundamental frequency, either in the stimulus or within the ear, is not a prerequisite. In this sense periodicity pitch is an auditory illusion.

It has also been suggested that periodicity pitch is due to distortion within the ear and that it has a physical basis in the existence of intermodulation products derived from components of the original stimulus. There is, however, ample evidence to show that this is not the case and that the presence of energy at the fundamental frequency, either in the stimulus or within the ear, is not a prerequisite. In this sense periodicity pitch is an auditory illusion.

It would be wrong to conclude that the pitch of a complex tone depends only on the frequency of the wave envelope; the fine structure of the waveform is also important. In fact, pitch bears a rather complicated relationship to the frequency spectrum of the stimulus and the shape and frequency of the wave envelope. An interesting and detailed review of this subject has been provided by Small (1970).

3.5 Difference Limens for Intensity and Frequency

Our ability to detect a change in the strength of a stimulus is the subject of a general psychophysical law due to Weber (1834): 'The increase in stimulus needed to produce the minimum perceptible increase in sensation is proportional to the pre-existing stimulus.' Weber's hypothesis can be written algebraically as

$$\Delta I/I = k , \qquad (3.3)$$

where I and $I + \Delta I$ are the physical magnitudes of two stimuli which are just perceptibly different, and k is a constant. Fechner (1860) proposed that k should represent an increment ΔS in sensation so that

$$\Delta S = k' \Delta I / I. \tag{3.4}$$

He further suggested that this expression could be written in terms of infinitesimals and integrated to give a subjective scale of loudness:

$$S = k' \log I / I_0. \tag{3.5}$$

The Weber–Fechner doctrine was widely accepted for almost a century and there is scarcely a text on experimental psychology that does not mention it. In recent times, however, it has been seriously challenged by Stevens. Equation (3.4) is clearly incompatible with the power law (equation (3.2)) although Weber's original concept (equation (3.3)) is not.

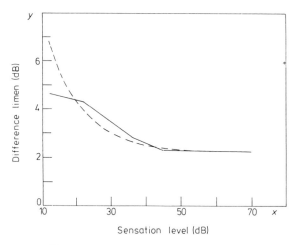

Sensation level (dB)

Figure 3.5 Difference limens for intensity. The ordinate is the smallest increment ΔI in intensity I that can be detected. This is expressed in dB; thus $y = 10 \log (I + \Delta I)/I$. The abscissa is the sensation level corresponding to the intensity I, that is, $x = 10 \log I/I_0$, where I_0 is the intensity at the threshold of hearing. There is no systematic variation of the difference limen with frequency; the full line is an average for the frequencies 128, 256, 440, 540 and 1000 Hz. The data can be represented approximately by the equation $y = 2 \cdot 3 + 15 \exp(-0 \cdot 1x)$ shown by the broken line. After Dimmick and Olson (1941). Reproduced with permission of the American Institute of Physics.

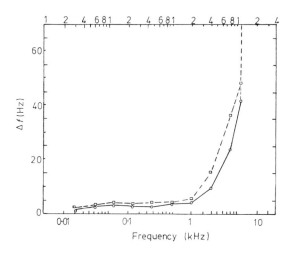

Figure 3.6   Difference limens for frequency at 15 dB (broken line) and 40 dB (full line) above threshold. Δf is the least discernible change in frequency of a tone whose frequency was modulated smoothly and continuously twice per second (Shower and Biddulph 1931). The nearly vertical line at the upper end leads to $\Delta f = 187$ Hz at 15 kHz (Wever and Wedell 1941). From Wever (1949).

The difference limen $\Delta I/I$ for acoustic intensity has been measured by a number of workers (see Wever 1949, Littler 1965). The results depend to a significant extent on the way in which the change in intensity is accomplished. Dimmick and Olson (1941) presented their subjects with two tones separated by a short silent interval. The least detectable increase in intensity was 4·6 dB near threshold falling to 2.3 dB at 70 dB above threshold. For sensation levels in the range 45–70 dB the difference limen was constant in accordance with Weber's law (figure 3.5).

Figure 3.6 shows difference limens for pitch obtained by Shower and Biddulph (1931). Frequency discrimination is practically independent of intensity at levels 15 dB or greater above threshold, but not surprisingly the limen increases at lower intensities as the threshold is approached. For the frequencies below 2 kHz the minimum preceptible change is approximately constant at 3–4 Hz but at higher frequencies it increases rapidly. The rising portion of the curve from 2 to 12 kHz corresponds to a constant relative difference $\Delta f/f$, approximately equal to 0·003 or 1/20 semitone.

3.6 Masking

Masking is a process whereby one sound is rendered inaudible by the presence of another. The most basic form of masking is perhaps the obliteration of a signal at threshold by physiological noise.

Mayer (1876) discovered that a pure tone could readily be masked by a second tone at a lower frequency, but not by one at a higher frequency. This asymmetrical behaviour is illustrated in figure 3.7 which shows how the threshold of audibility of a pure tone is raised when a second (masking) tone is present (Wegel and Lane 1924). The asymmetry, which is less marked when the masking tone has a low intensity or when its frequency is high, is also seen in the mechanical and neural tuning curves for the cochlea (figures 2.13 and 2.14). The notching of the masked threshold is due to the appearance of beats which assist detection of the signal. Wegel and Lane believed that the beats were generated by combination tones resulting from distortion within the ear. If the masking sound is a narrow-band noise, the notching is absent (figure 3.8).

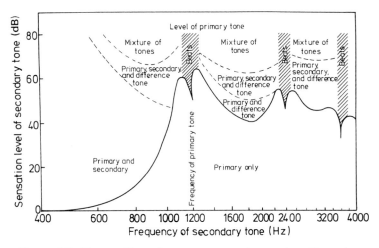

Figure 3.7 The masking of one pure tone by another. The primary (masking) tone is a sinusoid at 1·2 kHz presented 80 dB above threshold. The ordinate shows the level above the normal unmasked threshold at which a second tone becomes audible. The sensation produced by the combination of the two tones is indicated. After Wegel and Lane (1924).
Reproduced with permission of The American Institute of Physics.

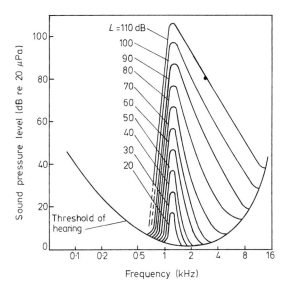

Figure 3.8 The masking of a pure tone by a narrow band of noise centred on 1·2 kHz. The ordinate shows the threshold sound pressure level of the tone in the presence of the noise. The parameter is the overall sound pressure level of the noise. The curve forming the base of the diagram is the unmasked threshold. After Zwicker 1958 *Acustica* **8** 237–58.

As described above, the masking sound and the signal being masked are present simultaneously. Masking can also occur when the masking noise is presented a little before or a little after the signal. The former, known as precedent or forward masking, is an example of auditory fatigue. This probably has its origin within the organ of Corti and represents a temporary loss of sensitivity following recent stimulation. Post-stimulatory or backward masking is more difficult to explain. According to Jeffress (1970) it is due to a late interaction between the masker and the signal in neural centres beyond the cochlea.

A remarkable feature of the auditory system is that the detection of a weak signal in one ear is virtually unaffected by the presence of a much louder sound confined to the contralateral ear. This is subject to the limitation that the threshold for the acoustic reflex is not exceeded and that the frequency of the sound in the contralateral ear is not equal or nearly equal to that of the signal (Ingham 1959). The ability of ears to act independently is particularly important in diagnostic audiometry since it permits a threshold to be determined in one ear while hearing in the other is masked out.

3.7 Critical Bands

The analysis of auditory phenomena is often simplified if the ear is imagined to possess a series of filters whose pass bands are known as critical bands. It should be understood that these filters are a convenient fiction created as an aid to analysis and description of auditory function. They should not be identified with real mechanical or neural elements. A critical band may be defined empirically as a range of frequencies within which subjective responses are constant, while at the boundaries of the band these responses undergo an abrupt change. For example, the loudness of a narrow band of noise at constant sound pressure remains unchanged as the bandwidth is increased until the critical value is reached. At this point a further increase in bandwidth is accompanied by a change (usually an increase) in loudness (figure 3.9). The term 'critical band' was originally used by Fletcher and Munsen (1937) to denote the band of frequencies within a masking noise that alone contributed to the masking of a pure tone at the centre frequency. Thus the threshold for a pure tone heard against a background of noise having constant energy per cycle rises as the bandwidth of the noise is increased up to the critical value, but thereafter extension of the noise is ineffectual. Critical bandwidths have been obtained from masking experiments employing this principle (Greenwood 1961). An alternative approach is to determine the pure tone thresholds in the presence of a uniform wide-band noise. This yields the critical ratio, defined as the ratio of the threshold intensity T of the pure tone to the spectral intensity per cycle L of the noise. Although this ratio is a frequency it is often expressed as the difference in the decibel values of T and L. In the presence of noise the pure tone becomes audible once the limiting signal-to-noise ratio α within the critical band is exceeded. The bandwidth Δf is therefore equal to $T/\alpha L$. When bandwidths given by this relationship are compared with values obtained by direct methods, it appears that α is independent of frequency and intensity throughout much of the auditory range (Scharf 1970). For frequencies above 200 Hz and overall noise levels from 20 to 90 dB SPL its value is 0·4. Detection of a pure tone in wide-band noise therefore requires

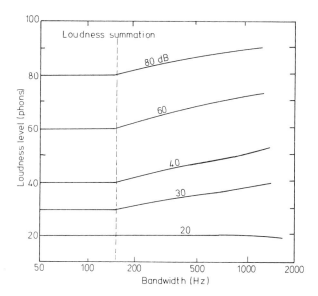

Figure 3.9 The loudness of a band of noise centred at 1 kHz as a function of bandwidth. The broken line shows the critical bandwidth at which the slope of the loudness function changes abruptly. For each of the five loudness curves the overall sound pressure of the noise is maintained constant as the bandwidth is varied. The parameter is the 'effective' level of the noise, that is, the amount of masking produced by one critical band. From Scharf 1970 *Foundations of Modern Auditory Theory* vol. I (New York: Academic Press).

a signal-to-noise ratio better than −4 dB for components of the noise within the appropriate critical band. A very much less favourable ratio can be tolerated with respect to components outside the band, particularly if they have a higher frequency than the signal (see figure 3.8).

The variation of critical bandwidth with frequency is shown in figure 3.10. This function is similar to the difference limen for frequency and both are directly related to the sharpness of the mechanical tuning in the cochlea. The boundaries of a critical band have a frequency separation of approximately 60 difference limens, corresponding to a change of 1·3 mm in the position of the amplitude maximum on the basilar membrane. Typical values of centre frequency and bandwidth of critical bands are listed in table 3.1.

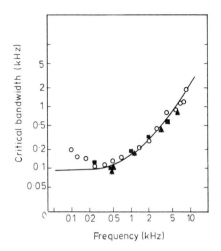

Figure 3.10 Critical bandwidth as a function of the frequency at the centre of the band. Data are from the following sources: —Feldtkeller and Zwicker 1956 (two-tone masking, phase sensitivity, loudness summation threshold); ▲ Greenwood 1961 (masking); ○ Hawkins and Stevens 1950 (2·5× critical ratio); ■ after Scharf (two-tone masking, loudness summation) 1970 *Foundations of Modern Auditory Theory* vol. I (New York: Academic Press).

3.8 Binaural Hearing

The advantages of binaural hearing are twofold: it provides a means for the location of the source of a sound, and it improves the discrimination of a particular sound of interest to the listener in the presence of surrounding noise. The latter is known as the 'cocktail party effect', it refers to our ability to follow at will the voice of one speaker when many others are talking simultaneously.

3.8.1 Location

There are two important interaural differences which enable the direction of a wavefront to be recognised. These are the difference in the time of arrival of the sound at the two ears, and the interaural difference in intensity, due mainly to the diffraction by the head. These differences are shown as a function of azimuth† in figures 3.11 and

† Azimuth is the angle in the horizontal plane specifying the direction of the source. The azimuth for a source directly in front of the observer is 0 degrees.

3.12. The broken line in figure 3.11 is the interaural time difference at two diametrically opposite points on the surface of a sphere of radius 87·5 mm representing the head. For a distant source this time difference is given by

$$\Delta t = rc(\theta + \sin \theta), \qquad (3.6)$$

where r is the radius of the sphere, c is the velocity of sound (343 m s^{-1}) and θ is the azimuth (Mills 1972). This equation is clearly an accurate description of the experimental data for the normal human head. The interaural intensity difference, on the other hand, is a complicated function of frequency and azimuth. It is negligible for low frequencies but may be as much as 20 dB at high frequencies.

Table 3.1 Centre frequency and bandwidth of the critical bands. From Scharf (1970) *Foundations of Modern Auditory Theory* vol. 1 (New York: Academic Press)

Number	Center frequency (Hz)	Critical band (Hz)	Lower cut-off frequency (Hz)	Upper cut-off frequency (Hz)
1	50	—	—	100
2	150	100	100	200
3	250	100	200	300
4	350	100	300	400
5	450	110	400	510
6	570	120	510	630
7	700	140	630	770
8	840	150	770	920
9	1 000	160	920	1080
10	1 170	190	1 080	1 270
11	1 370	210	1 270	1 480
12	1 600	240	1 480	1 720
13	1 850	280	1 720	2 000
14	2 150	320	2 000	2 320
15	2 500	380	2 320	2 700
16	2 900	450	2 700	3 150
17	3 400	550	3 150	3 700
18	4 000	700	3 700	4 400
19	4 800	900	4 400	5 300
20	5 800	1 100	5 300	6 400
21	7 000	1 300	6 400	7 700
22	8 500	1 800	7 700	9 500
23	10 500	2 500	9 500	12 000
24	13 500	3 500	12 000	15 500

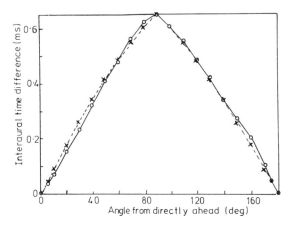

Figure 3.11 Interaural time difference as a function of azimuth for a source emitting clicks. —×— calculated values for a sphere (Woodworth 1938); —○— measured values for five adult male subjects (Feddersen *et al* 1957). Reproduced with permission of The American Institute of Physics.

The ability of a listener to locate a source can be expressed in terms of the minimum audible angle, defined as the smallest change in azimuth that can be detected. The method used by Mills (1958) for determining this angle was to have the observer seated in an anechoic room with his head in a restraining clamp. Tone pulses were emitted by a source which could be moved during a silent interval between two presentations of the tone. The observer then reported whether the second pulse came from the right or the left of the first. The smallest change in azimuth for which the direction of change could be correctly identified in 75% of a number of trials was taken to be the minimum audible angle (figure 3.13). At low and high frequencies small changes in azimuth could be detected, particularly when the source was straight ahead of the observer, but in the range 1·5–2 kHz the minimum audible angle became indeterminately large when the azimuth was greater than 45 degrees.

An alternative to using actual sources is to present sounds through earphones. In this way interaural differences in time, phase or intensity can easily be manipulated. The presentation is said to be *diotic* if the stimuli reaching the two ears are identical in all respects and *dichotic* if they are not. Sound delivered dichotically creates the illusion that it comes from within the observer's head rather than from his external environment, but lateralisation is still possible.

The bold curves in figure 3.14 show the minimum interaural phase difference detectable in a pair of equal-intensity tones presented dichotically (Zwislocki and Feldman 1956) and the minimum interaural intensity difference when the tones are homophasic. The lighter curves represent the corresponding quantities for a real source at an azimuth of one minimum audible angle. For frequencies up to 1·5 kHz the two phase functions are almost identical, showing that localisation in this frequency range is determined by interaural phase or time difference. From 1·5 to 6 kHz intensity difference appears to be the dominant factor, while at higher frequencies some other mechanism is operating.

Sounds in front of the observer are difficult to distinguish from sounds behind him. At frequencies below 6 kHz a source at azimuth θ produces almost the same interaural differences as a source at $\pi - \theta$.

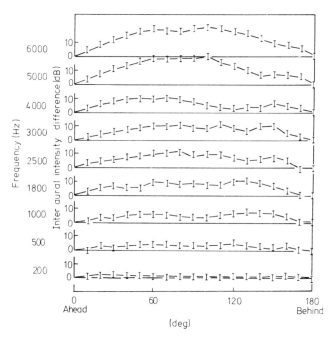

Figure 3.12 Interaural intensity difference as a function of azimuth and frequency. From Feddersen *et al* (1957). Reproduced with permission of The American Institute of Physics.

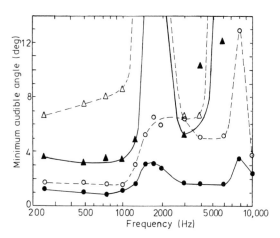

Figure 3.13 Minimum audible angle for tone pulses as a function of azimuth (● 0°; ○ 30°; ▲ 60°; △ 75°). From Mills (1972).

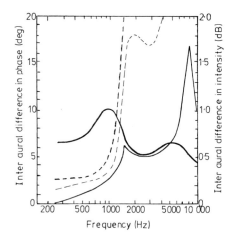

Figure 3.14 Comparison between the least detectable phase and intensity difference for dichotic (earphone) listening and the same quantities calculated for a real pure tone source at an azimuth of one minimum audible angle. For pure tones, time and phase are equivalent. The functions shown in this diagram would have the same shape if, instead of phase, interaural time difference had been represented on the ordinate. Phase: ––––– dichotic; ––––– real source. Intensity: –––––– dichotic; –––––– real source. From Mills (1972).

Even if this ambiguity is removed (the observer might be told that the source will always be in front of him) there are still difficulties. The location of a low-frequency source depends on the interaural phase difference but this may not provide sufficient information to specify the direction of the source uniquely. If ϕ is a positive phase difference (phase lead) associated with a given source, the direction of the source is uniquely specified only when

(*a*) $\phi_{90} < \pi$,
(*b*) $\pi < \phi_{90} < 2\pi$ and $\phi \le \pi - \phi_{90}$,

where ϕ_{90} is the maximum phase lead (azimuth 90 degrees) at the frequency concerned. Condition (*b*) applies for frequencies between 750 and 1500 Hz and above this frequency location on the basis of phase is always ambiguous. At higher frequencies, intensity differences are important. It can be seen from figure 3.12 that the rate of change of intensity difference with azimuth is very small for angles above 60° and this explains why it is difficult to locate a high-frequency source under these conditions.

Outside the laboratory the auditory input is seldom a sinusoid but consists largely of transients which provide timing cues that facilitate location of the source. According to Littler (1965) experiments with 'single sharp sounds' yield a minimum audible angle of 4°, corresponding to an interaural time difference of 60 μs. Experiments with transients are complicated by the fact that any stimulus has a finite duration in which it may be possible for the auditory system to extract on-going time or phase information which is unrelated to the transient interaural time difference associated with the onset of the stimulus. When care is taken to avoid transients (by switching tones on and off slowly) it is found that very small on-going time differences can be appreciated. For example, at 1 kHz Zwislocki and Feldman (1956) found that disparities of only 15 μs can be detected (this value can be obtained from the dichotic phase difference shown in figure 3.14). Even smaller values were found by Tobias and Zerlin (1959). They presented their subjects with bursts of filtered noise of various durations from 10 to 1000 ms and determined for each duration the least detectable interaural time difference. For tone bursts lasting more than 700 ms the interaural difference was independent of duration, having the amazingly small value of 6 μs. For shorter durations the time disparity increased at a rate of 2 μs for each halving of the duration. These results are of great importance because they show that the auditory system is capable of a highly refined temporal analysis which does not involve the place of maximum stimulation in the organ of Corti.

3.8.2 Binaural signal detection

Our ability to recognise the existence of a weak signal in the presence of noise is enhanced when listening with two ears rather than one. Suppose that the same pure tone signal is presented through earphones to both ears, and that the same masking noise is added to both channels until the tone becomes inaudible. If now a phase change is introduced into one of the signal channels so that the signals at the two ears are in antiphase, the audibility of the tone is restored. For example, using a tone at 500 Hz, Hirsh (1948) found that the antiphasic condition gave thresholds which were 11 dB better than those obtained in the homophasic condition. This phenomenon has been discussed by Jeffress (1972); following Webster (1951) he assumed that the noise in the critical band masking the signal could be treated as a sinusoid having a slowly varying phase and amplitude. Adding the tone to this gives a resultant whose phase and amplitude differ from that of the original noise. When signals are presented dichotically in antiphase two such resultants can be drawn, one for each ear, and a significant phase difference exists between them. It is suggested that the corresponding interaural time difference provides the basis for the detection of the signal.

3.9 Auditory Analysis of Complex Sounds

Some 20 years after the publication of Fourier's famous mathematical analysis, Ohm (1843) proposed that the ear could reduce a complex waveform to its sinusoidal components and that these components alone gave rise to the sensation of hearing. This principle was developed further by Helmholtz (1863) who suggested that nonlinearity within the ear provided the appropriate sinusoids when these were lacking in the external stimulus (as for example, in Seebeck's experiment). Helmholtz also believed that the basilar membrane acted as a series of resonators, each of which stimulated a nerve or small group of nerves specific for the frequency concerned. His hypothesis was therefore one of the 'place' theories of hearing. Such theories allege that different frequencies are represented separately in different regions of the auditory nervous system. The perception of frequency is therefore dependent on the spatial distribution of neural events and not on the transmission properties of individual nerve fibres. Thus the ability of the ear to respond to frequencies above the limit for synchronous neural activity is explained. The chief problem with Helmholtz's theory was that the resonators had to be heavily damped in order to comply with the obvious fact that the sensation of hearing does not persist after the stimulus has been removed; in fact, rapid changes in intensity are readily discernible. Heavy damping of the resonators was inconsistent with their requirement for high selectivity commensurate with the fine pitch discrimination of the auditory system. The problem was partly solved by Békésy's discovery of the travelling wave mechanism. This removed the need to postulate a series of resonators while at the same time the place principle could be retained. In Békésy's model, frequency discrimination is explained by a neural process which restricts transmission to those neurones innervating the region of maximum stimulation while elsewhere the response is inhibited.

The place theory cannot readily explain the existence of periodicity pitch since it has been shown that this is not simply a consequence of nonlinear distortion. The perception of periodicity might therefore be the result of a central process which does not involve the frequency-specific properties of individual nerve fibres. This explanation is an example of the 'telephone' theory (originally due to Rutherford 1886) in which the analytical process is relegated to higher centres within the brain. Wever (1949) proposed that both place and telephone theories might be appropriate depending on the frequency involved. Thus the perception of high-frequency components is dependent on the place of maximum stimulation while the analysis of low frequencies, for which the vibratory pattern on the cochlear partition does not possess well defined maxima, is based on 'volleys' of neural impulses which are produced by the simultaneous firing of many neurones and which carry the periodicity of the original stimulus.

The most well defined and easily produced artificial stimulus is the sinusoid. It is therefore not surprising that the pure tone has been the most commonly used stimulus for the investigation of auditory function. As a consequence, auditory theory has been preoccupied with the concepts of frequency and phase—terms which are generally meaningless except with regard to recurring quantities. Thus the cochlea is usually regarded as a frequency analyser because individual neurones or specific regions of the basilar membrane are found to be selective in their frequency response. While this is an undoubted experimental fact, it does not follow that frequency analysis as such is the physiological function involved. The frequency-dependent properties might well be fortuitous. Nordmark (1970) virtually rejects the place theory as an explanation of the analysis of complex sounds and he considers that auditory theory has devoted relatively little attention to the problem. As the eye perceives patterns in space, so the ear perceives patterns in time. According to Nordmark the ear should be regarded as a temporal pattern analyser. 'Once this notion is accepted,' he writes, 'I am convinced that the phenomena of auditory analysis will turn out to be considerably easier to explain than is at present thought possible.'

4 Electrophysiology

4.1 Introduction

A study of electrical activity in and around the cochlea is a valuable method of investigating auditory function. In recent years this method has been extended to include electroencephalic responses that follow auditory stimulation. The phenomena to be described in this chapter are directly related to the electrical activity within the neurones of the auditory system and cerebral cortex. While most of these phenomena are the result of events occurring simultaneously in many neurones, the logical starting point is an account of the processes occurring in a single nerve cell. The description which follows is necessarily simplified and brief. For further information the reader should refer to texts dealing with neurophysiology. A good introduction to this subject, together with a suitable bibliography, is provided by Stevens (1966) and Brindley (1974a).

4.2 Nerve Conduction

A nerve cell or neurone consists of a cell body and one or more thread-like processes emanating from it. The cell body contains the nucleus and organelles responsible for regulating the metabolic activity of the cell. The processes are called dendrites or axons, the distinction being that dendrites conduct signals towards the cell body, and the single axon conducts signals away from the cell body. Sensory nerve cells are called bipolar because they have a nerve fibre or dendrite at one end of the cell and an axon at the other end, and this is the arrangement found in the spiral ganglion of the cochlea and also in the vestibular ganglion of the vestibular division of the VIII nerve. In most sensory nerves, however, these two processes have converged so that they meet at one side of the cell body to form a T-shaped junction. One arm of the 'T' is the dendrite from the peripheral sense organ, and the other is the axon proceeding centrally, but the appearance is of one continuous, thread-like nerve fibre with the cell body attached to it by

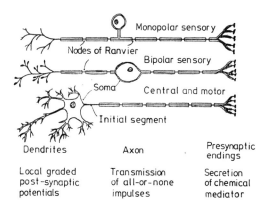

Figure 4.1 Three varieties of neurone, showing possible relations of the cell body (soma) to the axon. After Davis 1961 *Physiol. Rev.* **41** 391–416.

the vertical limb of the 'T' (figure 4.1). Nerve fibres may be sheathed in a fatty material called myelin which acts as an electrical insulator. In peripheral nerves the sheath is interrupted at intervals by constrictions known as the nodes of Ranvier. Both dendrites and axons terminate in twig-like branches sometimes called telodendrons. The terminal branches of axons end in loops or pre-synaptic endings (boutons terminaux) which form functional junctions known as synapses with the dendrites or cell body of another neurone (figure 4.2). The relation of two neurones at a synapse is one of near contact but not continuity of substance; adjacent surfaces of the neurones in a synapse are separated by a small gap (0·02–0·03 μm) known as the synaptic cleft. Information transmitted in an axon passes from its pre-synaptic terminals across the synapse to the dendrites or body of the following neurone. Since there are numerous connections between neurones, each receives information from many sources. This information is 'integrated' (spatial and temporal summation) before being relayed to succeeding neurones. The information carried by an axon is in the form of an impulse and the associated electrical manifestation is a voltage pulse known as the action potential. Impulses arriving at the axon terminals cause the release of a chemical transmitter which rapidly diffuses across the synapse and modifies the electrochemical activity in the receiving dendrite. According to the characteristics of the particular synapse involved, the post-synaptic activity is either increased or decreased by the arrival of the chemical transmitter.

When the transmitter is excitatory the dendritic potential rises and, if sufficiently large, it initiates an impulse in the receiving axon.

In common with most animal cells, the interior of a neurone has a different electrolytic composition from that of the extracellular fluid. This is because the cell membrane (plasma membrane) is not equally permeable to all ions, and because the sodium ion is actively transported across the membrane by a metabolic process known as the 'sodium pump.' In the extracellular fluid the principal ions are Na^+ and Cl^-, whereas in the cell protoplasm they are K^+ and negative organic ions. In its resting state the cell membrane of a neurone has a low permeability to sodium and the entry of this ion is compensated by the action of the sodium pump. The interior of the cell is therefore deficient in sodium and thus it acquires a negative charge by virtue of the organic anions. This charge is almost neutralised by the influx of potassium, for although the cell membrane hinders the passage of this

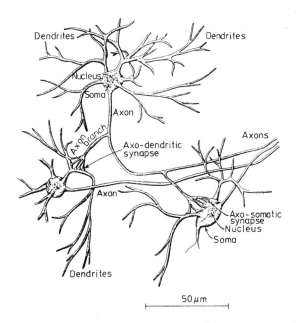

Figure 4.2 Semi-schematic representation of three neurones showing possible interconnections. After Stevens *Neurophysiology* Copyright © 1966. Reprinted by permission of John Wiley and Sons, Inc.

ion it is by no means impermeable to it. Consequently the concentration of potassium inside the cell is large compared to the concentration outside it. Diffusion tends to produce a net efflux of potassium so that a residual negative charge inside the cell is needed to maintain the concentration gradient. The potential on the inside of the cell membrane, relative to the potential of the extracellular fluid, is known as the membrane potential. If only one type of ion is considered, the resting value of this potential is given approximately by the Nernst equation:

$$E = (RT/nF) \ln [K^+]_o/[K^+]_i, \qquad (4.1)$$

where R is the gas constant, T is the absolute temperature, F is the Faraday and n is the valency of the ion (unity).

The permeability of the cell membrane of an axon depends on the membrane potential, with the result that the membrane polarisation is metastable. This provides an explanation of the generation and transmission of an impulse in an axon. If the membrane potential at some point on an axon is increased as a result of stimulation the permeability of sodium is also increased in the stimulated region. This allows a local influx of positive ions which raises the potential still further. The situation is therefore unstable and leads to a rapidly rising potential (action potential) shown in figure 4.3. During this phase the permeability of sodium becomes large compared with the permeability of potassium, and the sodium ions enter in sufficient numbers to make the interior of the cell positive. The peak of the action potential is therefore determined principally by the sodium concentration gradient. The permeability to sodium does not remain high but shortly before the peak of the action potential is reached it starts to fall (a process known as sodium inactivation). This is followed by a temporary increase in potassium permeability. As a result, the membrane potential is restored to its resting value.

Transmission in an axon is normally directed towards the synaptic terminal. While the nerve is conducting, a depolarised region with its associated impulse travels along the axon. At the boundary of this region ahead of the impulse further depolarisation is occurring, while behind the impulse recovery is taking place. In the recovery phase there is a period known as the absolute refractory period during which no stimulus, however strong, can produce a further impulse. This is followed by the so-called relative refractory period in which an impulse can be initiated if the stimulus is sufficiently intense. If the cell is in its

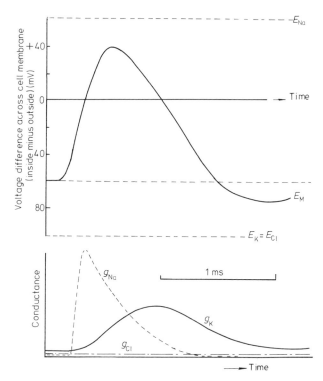

Figure 4.3 Changes in membrane potential and permeability at a point on an axon during the conduction of a single impulse. E_m is the membrane potential (upler graph) and E_{Na}, E_K and E_{Cl} are the equilibrium potentials for each of the ions taken singly. The conductances (reciprocal permeability) of the ions are shown in the lower graph. After Roberts (1966).

resting state an impulse is initiated only if the stimulus exceeds a definite threshold value characteristic of the neurone concerned. The associated action potential is, however, independent of the magnitude of the stimulus—a feature known as the 'all-or-none' property of nerve conduction. During the relative refractory period an impulse can be produced but the threshold for stimulation is higher than normal, and in this case the response does depend on the strength of the stimulus. The duration of the cycle shown in figure 4.3 is about 1 ms so that the maximum rate at which a single fibre can conduct is approximately 1000 pulses per second. The velocity of propagation of the impulse ranges from 0·1 to 100 m s^{-1} depending on the fibre concerned. It is more rapid if the axon is surrounded by an insulating layer of myelin. In this case the conduction occurs as a series of 'jumps' between the nodes of Ranvier.

If the potential at a *point* on the membrane of a neurone is changed by the influence of some local process, the potential at neighbouring points is also changed. This is known as passive (or electronic) spread of the change in membrane potential. The change is said to be *passive* because it does not involve metabolic activity or alterations in the membrane permeability. As might be expected, the passive change in potential diminishes with increasing distance from the point of origin. In a non-myelinated fibre, depolarisation beyond the boundary of the active region is accomplished by passive spread so that the impulse is propagated as described. In a myelinated axon, the depolarisation at one node is communicated passively to the next where it is reinforced by local action.

The plasma membranes of dendrites and cell bodies, unlike those of the axons, do not alter their permeabilities in response to changes in potential. They are therefore electrically stable and cannot transmit information in the same way as axons. Changes in permeability do, however, accompany the excitation or inhibition produced by chemical transmitters received from the axon terminals. These changes are selective for particular ions and result in an increase or decrease in the membrane potential. The potential in a dendrite or cell body may be called 'post-synaptic' to distinguish it from the potential in an axon. Compared with the action potential, the post-synaptic potential has a long duration and so may also be called a 'slow' potential†. Post-synaptic potentials are graded according to the amount of chemical transmitter received and spread passively to the axons. If the potential is positive and above threshold, it will initiate an impulse in the axon. Moreover, some axons continue to fire repeatedly as long as a sufficient dendritic potential is maintained. The reason for this is not clear and a satisfactory explanation is unfortunately not contained in the discussion so far. The repetition frequency depends on the magnitude of the potential so that a pulse-coding system exists. Decoding takes place by integration of the 'bursts' of chemical activity in the synapse receiving the pulse train. The importance of this in audition is that the

---

† This should not be confused with the slow evoked vertex potential although the latter is dendritic in origin.

frequency of the impulses carried in a single neurone is not necessarily equal to the frequency of the acoustic stimulus.

Because of the multiple connections between neurones, a given neurone always receives a large number of excitatory inputs and is thereby maintained in a continuous state of 'spontaneous' activity. This activity is modified (enhanced or reduced) according to the potential in any one dendrite. Thus all inputs to the neurone are able to influence its behaviour. But for this, inhibitory inputs would be ineffective because there would be no activity to inhibit and consequently information would be lost.

4.3 Measured Potentials: Near- and Far-field Recording

The cell bodies of neurones are typically some $30\ \mu$m across and the diameter of an axon is considerably less. Because these cells are small and often anatomically inaccessible it is not usually possible to observe membrane potentials directly. Despite the technical difficulties, events in single auditory neurones have been detected using microelectrodes, but this type of recording is the exception rather than the rule. Much of the experimental work that led to the theory of nerve conduction was done using an exceptionally large neurone found in the squid. This nerve has an axon about $500\ \mu$m in diameter so that it is relatively easy to put electrodes into the interior of the cell.

Recording techniques are classified as *near field* if the electrode is in or close to the electrically active tissue, and *far field* if it is not. Electrodes in or close to a nerve trunk containing many fibres—such as the auditory nerve—record near-field potentials representing synchronous activity in a number of the individual fibres. Processes within the cochlea also create near-field potentials observable with electrodes inside or close to the cochlea. Useful sites for the electrode are the promontory, which is the bulge in the wall of the tympanic cavity created by the basal turn of the cochlea, and the round window niche. On the other hand, events in the brainstem and cerebral cortex are usually recorded with electrodes at a considerable distance (perhaps several centimetres) from the source. Both near- and far-field techniques also require a reference electrode at a remote point whose potential relative to earth is unaffected by the processes under investigation. A good site for the reference electrode is the neck, but in some circumstances the mastoid process or the earlobe is preferable.

An extracellular electrode cannot give a direct indication of the membrane potential. Instead it responds to potentials generated in the extracellular medium by the flow of charged particles into and out of the neurone. Both near- and far-field responses reflect activity in many neurones but the near-field response is strongly influenced by events close to the electrode. The far-field effect is related to processes occurring in a relatively large volume. The activity must be distributed in such a way that contributions from the elements contained in the volume do not cancel at the recording electrode.

As a basis for analysis, the extracellular medium is usually treated as a homogeneous three-dimensional (volume) conductor. The potentials due to currents in the conductor can then be represented in an equivalent electrostatic model in which the conducting volume is replaced by a uniform dielectric. Each neurone is represented by a hollow insulator whose inner and outer surfaces are oppositely charged to form a dipole layer with a strength proportional to the membrane potential. Alternatively, the neurone can be represented by a solid insulator with a surface charge proportional to the normal component of the membrane current. The far-field potential at the site of the electrode has then to be calculated by summing the potentials due to all neurones in the model. An elementary account of this method of analysis is given by Brindley (1974b).

A point at which positive charge flows out of a neurone is sometimes called a source, and a point where it flows in is called a sink. In a volume of brain tissue containing a great many neurones there may be an extended source in one region and a corresponding sink in another. For this to happen the elements must be approximately aligned and their activities synchronised. Such an arrangement is believed to exist in parts of the cerebral cortex where the tissue contains a vast number of apical dendrites belonging to neurones known as the pyramidal cells. The dendrites are well oriented perpendicular to the outer surface of the brain. It is thought that this organisation is capable of producing a dipole corresponding to a source and a sink aligned normal to the surface of the cortex. Such dipoles are generated by slow post-synaptic processes rather than action potentials and are probably the origin of some of the evoked potentials recorded on the skull.

4.4 Resting Potentials in the Cochlea

Unless otherwise stated, all potentials refer to the potential of a remote electrode placed, for example, on the neck.

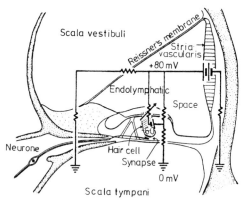

Figure 4.4 Equivalent circuit to represent electrical activity in the cochlea. The variable resistor performs the modulation of the receptor current in the hair cell in accordance with the mechanical stimulus. The batteries represent metabolic processes in the stria vascularis and hair cell, which generate resting potentials of +80 and −60 mV, respectively. Activity in the nerve fibres is not represented. From Davis and Silverman (1970) *Hearing and Deafness* © Holt, Rinehart and Winston.

In the absence of stimulation, quiescent potentials exist in the cochlea due to variations in electrolytic composition, selective differences in membrane permeability, and metabolic process such as the sodium pump already described. Nearest to body potential is that of the perilymph, but even this has a small positive voltage which varies with position on the spiral, reaching +5 mV at the apex (Békésy 1960). In contrast, the endolymph in the scala media has a relatively high positive potential (+80 mV) known as the endocochlear potential, while the hair cells have a negative potential of the same order (−60 to −80 mV). The metabolic energy needed to maintain these potentials is probably generated in the hair cells and the border cells of the stria vascularis and, according to a scheme suggested by Davis (1965), the remaining tissues are electrically passive (figure 4.4).

4.5   Evoked Potentials in the Cochlea

Evoked potentials are so named because they appear during or shortly after stimulation. Potentials evoked by auditory stimuli are obtained from the cochlea and central nervous system and may be classified according to their characteristic latency. The earliest responses occur, of course, in the cochlea and so come under the heading 'first' in table

Table 4.1 Auditory evoked potentials.

	Class	Probable source	Latency (ms)	Best response	ERA
E Coch G	First	Organ of Corti (CM=external hair cells)	0	SP (DC) ⎱ CM (AC) ⎰	?
		N VIII	1–4	AP (N_1)	* *
Potentials	Fast	N VIII, brain stem	2–12	P_6	* *
	Middle	Neurogenic: cortex I ⎱ Myogenic: 'sonomotor' ⎰	12–50	P_{35}	* ?
	Slow	Cortex II (waking)	50–300	$\{\ N_{90}-P_{180}-N_{250}$ sustained potential (DC)	* *
Vertex		Cortex III (asleep)	200–800	$P_{200}-N_{300}$: N $\xrightarrow{600}$ P	*
	Late	Cortex IV (expectation)	$\{$ 250–600 DC shift	P_{300} ⎱ CNV ⎰	?

Notes

(*a*) The symbols P and N denote positive and negative turning points in the waveform of the response where the evoked potential refers to a remote point such as the neck. The subscript indicates the characteristic latency of the maximum or minimum measured in ms from the appearance of the cochlear microphonic. Thus the response in the example is designated $P_a-N_b-P_c$. The latency of the onset of the cochlear microphonic with respect to the arrival of the stimulus is negligible.

(*b*) In an alternative notation the subscripts denote the positions of N and P in a sequence. Thus in the example, P_a is P_1, N_b is N_1, and P_c is P_2. The action potential shown as N_1 in the table is the first minimum in the response; its latency is usually greater than 1 ms.

(*c*) The responses shown in the table are for normal adult subjects with a stimulus presented at a sensation level of 70 dB.

(*d*) Entries in the column headed ERA indicate the usefulness of the response in electric response audiometry. From Davis 1976 *Ann. Oto-Rhino-Laryngol.* **85** suppl. 28.

4.1. A record of these potentials is called an electrocochleogram (ECochG or ECoG). The first responses are of three types: a DC response known as the summating potential (SP); an alternating response called the cochlear microphonic (CM); and a transient response due to gross action potentials (AP) from first-order neurones of the cochlear nerve.

Before describing these responses individually, it is appropriate to mention the origin of evoked electrical activity in the cochlea and the events leading to the production of action potentials in the auditory nerve. The fibres which innervate the organ of Corti are non-myelinated; they are dendrites of the bipolar sensory neurones which collectively form the acoustic nerve and which have their cell bodies in the spiral ganglion within the modiolus. The neurones are myelinated to the point known as the habenula perforata where they emerge from the osseous spiral lamina. The dendrites make synaptic junctions with the hair cells and, as mentioned in Chapter 2, the inner cells are densely innervated, each being associated with many afferent neurones. In the outer rows the hair cells are sparsely innervated and here the dendrites branch to make multiple connections with several cells (figure 4.5).

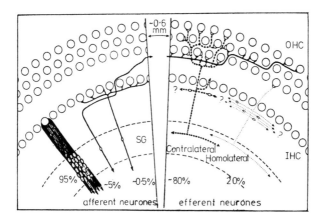

Figure 4.5 Innervation of the organ of Corti in a plane perpendicular to the modiolus. Different methods of innervation by afferent fibres (full lines) and efferent fibres (broken lines) are shown, together with the percentage of each type. OHC outer hair cells; IHC inner hair cells; SG spiral ganglion. The apex of the cochlea lies to the left of the diagram. After Spoendlin 1975 *Audiology* **14** 383–407.

It is thought that the electrical resistance or the potassium permeability of the reticular surface of a hair cell is altered when the stereocilia are stimulated mechanically. This modulation is represented by the variable resistance in figure 4.4. The change in resistance may correspond to displacement or possibly the rate of change of displacement (velocity) of the hairs. The entry of additional potassium raises the membrane (receptor) potential of the cell and so triggers the release of a chemical transmitter at the synapse. This in turn leads to the generation of post-synaptic potentials in the dendrites and to the production of action potentials in the axons of the auditory nerve. It is scarcely surprising that it has not been possible to make direct observations of the receptor and dendritic potentials associated with the action of individual hair cells; rather the existence of these potentials has been inferred from measurements in other branches of neurophysiology and from observations using intracochlear electrodes. The latter respond to a weighted average of contributions from many elements in the vicinity of the electrode.

4.5.1 The cochlear microphonic

The cochlear microphonic was discovered by Wever and Bray in 1930. It is an alternating potential which, as its name implies, follows the stimulus in the same manner as the output of a microphone. The microphonic response has a negligible latency and indeed its onset is often used as a reference from which the latencies of other responses can be determined.

It is generally accepted that the cochlear microphonic is generated by the hair cells themselves and thus it corresponds closely to the alternating component of the receptor current and the associated mechanical events in the organ of Corti. The cochlear microphonic is produced mainly by the outer hair cells. This has been demonstrated by experiments in which the outer cells were selectively destroyed by an ototoxic agent (Dallos 1975). There is also evidence suggesting the existence of a weak component attributable to the inner hair cells and it is thought that this is proportional to the derivative of the major component (Dallos *et al* 1972). For sensation levels up to 70 dB, the cochlear microphonic increases linearly with the stimulus but considerable distortion occurs at high levels (figure 4.6). This distortion does not necessarily reflect nonlinearity in the mechanical behaviour of the cochlear partition, but is probably for the most part electrical in origin.

The microphonic can be recorded at any point close to or inside the

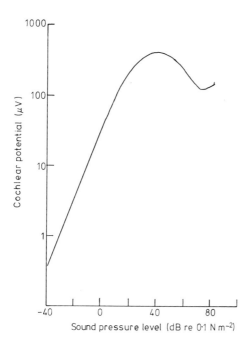

Figure 4.6 Cochlear microphonic recorded in the middle ear of an anaesthetised cat. The stimulus is a 1 kHz sinusoid and the acoustic pressure is measured in the meatus close to the eardrum. The recording was made in the course of experiments on distortion in the ear. From Wever and Lawrence 1954 *Physiological Acoustics* (reprinted by permission of Princeton University Press).

cochlea, but the response relates to 'local' events only when the vibration of the basilar membrane has its maximum in the neighbourhood of the electrode. The arrangement which provides the most specific information employs a pair of intracochlear electrodes at opposing sites in the scala vestibuli and scala tympani. As the equivalent circuit in figure 4.4 shows, alternating potentials associated with local events on opposite sides of the partition are in antiphase and can therefore be added by connecting electrodes to a differential amplifier. This technique has the merit that action potentials which would normally be present in the recording occur as common-mode signals and are rejected. The output is such that the polarity of the scala

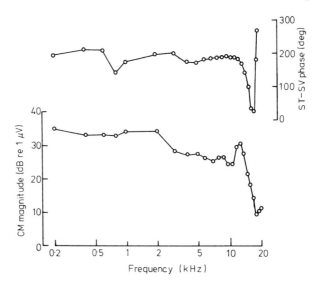

Figure 4.7 Cochlear microphonic recorded from a pair of electrodes in the basal turn of the cochlea of a guinea pig. The stimulus is a sinusoidal motion of the stapes with a constant rms velocity of 57 dB re 1 nm s^{-1}. The upper plot shows the phase difference of the potentials in the scala tympani and scala vestibuli. The lower plot shows the magnitude of the response. After Dallos 1975 *Human Communication and Its Disorders* (New York: Raven).

tympani is negative during the condensation phase of the acoustic stimulus.

As already stated, the response registered by a particular electrode is determined by both local and distant activity. This is illustrated in figure 4.7 which shows how the amplitude of the cochlear microphonic and the interelectrode phase difference vary with frequency. The electrodes are in the basal turn at a point where the amplitude of the travelling wave is a maximum at 13 kHz, a frequency called the 'best' or 'cut-off' frequency of the response. Below this frequency the vibration of the basilar membrane is fairly uniform in the region accessible to the electrodes; the rate of change of amplitude and phase with position on the membrane is small. Thus all events in the region of interest combine to give a response which corresponds to the activity at the site of the electrodes. Under these conditions the amplitude of the microphonic is fairly constant and the interelectrode phase difference is close to 180 degrees. Above the best frequency the opposite situation exists. The amplitude and phase of the vibration change rapidly with position so that significant differences exist between neighbouring parts of the recording field. Now the interelectrode phase changes dramatically with frequency, demonstrating that the recorded potentials no longer relate to local events. Nevertheless, the location of the maximum in the travelling wave corresponds accurately to the position of the electrodes if the best frequency of the response is used as an indicator (Dallos 1973, 1975).

The variation of the phase of the microphonic with frequency is shown in figure 4.8. In the diagram the phase† of the electrical response is referred to the velocity of the stapes. At low frequencies electrodes in the basal turn record a microphonic which is in phase with the velocity of the stapes. In these circumstances the displacement of the basilar membrane is, according to Békésy, in phase with the displacement of the stapes and therefore the velocities of the membrane and stapes are also in phase (see figure 2.12). This seemingly conflicts with the proposition (Dallos 1975) that the membrane displacement is the chief source of the microphonic, although it does not

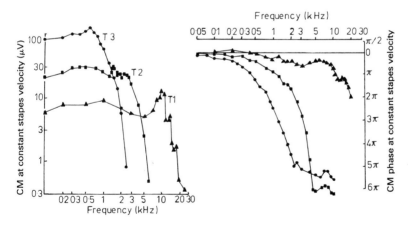

Figure 4.8 Amplitude and phase of the cochlear microphonic recorded in the first, second and third turns of the cochlea of a guinea pig. The stimulus is a sinusoidal motion of the stapes with a constant rms velocity. The phase of the response is referred to the velocity of the stapes. After Dallos 1973. *Basic Mechanisms of Hearing* (New York: Academic Press).

† This should not be confused with the interelectrode phase difference.

contradict the statement that these quantities are proportional to each other.

4.5.2 Summating potentials

The summating potentials are DC changes which accompany the cochlear microphonic. They are derived principally from receptor potentials in the organ of Corti but dendritic activity may also contribute.

Two types of summating potential can be recognised: the first is the potential difference between the scala vestibuli and the scala tympani; and the other is the average of these potentials. At a given recording site both components change systematically with frequency, as shown in figure 4.9. Below the best frequency the difference potentials are positive. They become negative in the vicinity of the best frequency and are negligible at higher frequencies. The average potential is positive in the neighbourhood of the best frequency and negative elsewhere.

When the frequency of the stimulus is near to the best frequency the summating potentials vary linearly with stimulus intensity (figure 4.10(a)). However, when the frequency is much higher, so that the

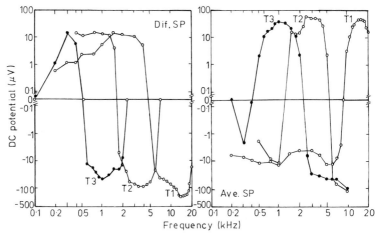

Figure 4.9 Magnitude of the summating potential recorded in the first, second and third turns of the cochlea of a guinea pig (three different animals). The stimulus is a sinusoidal motion of the stapes with a constant amplitude of 1 nm. *Left:* summating potential measured differentially (scala vestibuli–scala tympani). *Right:* average potential in the two scalae relative to a remote point. After Dallos 1975 *Human Communication and Its Disorders* (New York: Raven).

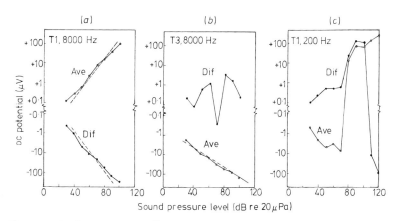

Figure 4.10 Average and differential summating potentials in the first and third turns of the cochlea of a guinea pig measured at 0·2 and 8 kHz as a function of sound pressure in the meatus. In (a) the recording electrodes are close to the place of maximum vibration on the basilar membrane. In (b) and (c) they are well removed from the maximum, being either nearer the apex (b), or base (c). After Dallos 1975 *Human Communication and Its Disorders* (New York: Raven)

electrodes are well removed apically from the position of the maximum in the travelling wave, the difference component is small (Dallos describes it as negligible). The average potential is then negative, falling linearly with increasing intensity (figure 4.10(b)). If the electrodes are placed basally with respect to the maximum (figure 4.10(c)) the difference potential is positive, rising somewhat irregularly with intensity. The average is at first negative (as in figure 4.9) but as the intensity is increased it becomes positive and, after passing through a maximum, reverts rapidly to a large negative value. The latter is possibly an indication of activity in the dendrites.

4.5.3   Action potentials in the auditory nerve: single neurones
Activity in the auditory nerve has already been mentioned in Chapter 2 in connection with mechanical and neural tuning mechanisms in the cochlea (figure 2.14), and a detailed account is to be found in the article by Evans (1975). Measurements obtained from single fibres in a cat's ear (Kiang 1965) show that the neurones of the auditory nerve are continuously active and, in the absence of acoustic stimulation, discharge spontaneously at rates ranging from about 5 to 100 impulses

per second. The discharges occur randomly and obey Poisson statistics. The average impulse rate is fairly constant for a given fibre, but it varies widely from one neurone to another and is apparently unrelated to the characteristic frequency (CF) at which the fibre is 'tuned.' In the presence of a sustained stimulus the average rate of discharge is increased and, once the threshold for the fibre has been exceeded, this increase is approximately proportional to the intensity of stimulation. Fibres whose characteristic frequencies are close to that of the stimulus are of course the most easily influenced. The maximum firing rate that can be evoked is in the range 100–200 impulses per second. It varies from one fibre to another but is generally greatest in units which have high rates of spontaneous activity. Discharges are also statistical events when a stimulus is present but in this condition the probability of their occurrence varies throughout the stimulus cycle. For frequencies up to at least 4 kHz (in a cat) there is a definite moment in each cycle when this probability is a maximum. It corresponds to the instant when the potential in the scala vestibuli changes from positive to negative with respect to the potential of scala tympani (Deatherage *et al* 1959). Thus there is a tendency, particularly at lower frequencies, for discharges to be synchronised with the stimulus—a process known as 'phase locking'.

If the level of the stimulus remains constant the average evoked activity in each cycle is also fairly constant, but a change in the stimulus initiates a large transient response. The sudden onset of an otherwise continuous tone can raise the discharge rate by as much as 500 impulses per second, but the response quickly adapts and falls asymptotically to the steady value. Similarly, when the tone is turned off the level of activity falls temporarily below the normal spontaneous discharge rate for the fibre. The transient increase with the appearance of the stimulus is known as the 'on-effect.' It has an important influence on the gross action potential of the nerve trunk and provides the basis for most forms of electrocochleography.

4.5.4 Gross action potentials in the auditory nerve

Simultaneous activity in a large number of axons in the auditory nerve produces a gross (compound) action potential that can be observed as a near-field response. Before describing this response it is necessary to mention the methods of recording it and the type of stimulus commonly used.

A favourite site for the recording electrode is the promontory, but the round window, the tympanic annulus and the external ear are also used. The potentials to be observed are small compared with the cochlear microphonic and much smaller than action potentials obtained from individual fibres. For example, a transient stimulus (click) presented at a sensation level of 90 dB produces an action potential of only $10 \mu V$ at the promontory (Yoshie and Ohashi 1969). Near threshold the response is only about $0 \cdot 1 \mu V$, a level close to the limit of detectability in a practical recording system. In order to detect these weak signals it is necessary to average a large number of responses derived from repeated presentations of the stimulus. The practical requirements of averaging preclude the use of any stimulus other than a transient and, moreover, the response is itself evanescent even when stimulation is prolonged. The best stimulus is a short tone burst, though clicks or filtered clicks are more commonly used. Although the power spectrum of a transient signal is spread over a wide range of frequencies it is possible to shape the envelope of a tone burst in such a way that most of the energy appears at the frequency of the tone. The advantage of this type of stimulus is that with it the frequency dependence of the response can be investigated and in diagnostic use auditory thresholds can be determined at each of several frequencies. Eggermont *et al* (1974) describe in detail the method of generating the tone burst and the techniques for amplifying and averaging the response. In their system, alternate tone bursts are presented in opposite phase in order to cancel the cochlear microphonic.

Action potentials in the electrocochleogram represent brief episodes of coordinated activity in the auditory nerve. The gross response to each cycle of the tone is diphasic and is usually dominated by an initial negative wave (N_1) associated with the onset of the stimulus as illustrated in figure 4.11. The waveform shown in this diagram should not be confused with that of the membrane potential (figure 4.3). An electrode in the middle ear is effectively connected via the perilymph to the afferent end of the VIII nerve. The initial fall in cochlear potential is associated with the flow of positive ions into the nerve during the rising phase of the membrane potential. Figure 4.11 shows the response obtained to tone bursts of various frequencies. In this example the stimulus has a trapezoidal envelope consisting of six periods at constant amplitude and rise and fall times of two periods each. Thus the overall duration of the stimulus varies inversely with frequency. At the lowest frequencies, small fluctuations in the response can be seen corresponding to neural discharges synchronised with each cycle of the stimulus (the apparent frequency doubling is due to the

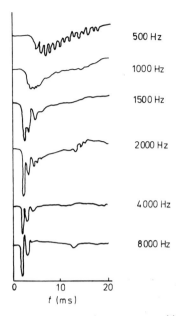

Figure 4.11 Electrocochleograms for a person with normal hearing.
For explanation see text. From Eggermont *et al* (1974).

phase reversal between successive stimuli). As the frequency is increased, adaption and refractoriness become increasingly important,
with the result that activity in individual units is no longer seen as a
coordinated response. The only event to be detected is then an initial
burst of synchronous activity at the onset of the stimulus, and it is this
that produces the first negative deflection shown in the diagram. It is
also noticeable that the latency of N_1 is inversely related to the
stimulus frequency. This is due to the progressive displacement of the
maximum in the Békésy wave towards the base of the cochlea and a
corresponding reduction in the time taken for effective transmission of
the acoustic signal in the inner ear.

The response at a fixed frequency (2 kHz) is shown in figure 4.12 as
a function of stimulus intensity. At intermediate intensities (65 and
75 dB) the initial wave is split into two components, $N_1(I)$ and $N_1(II)$,
separated by an interval of approximately 1 ms. Only the first of these
components is evident at high intensities and only the second at low

intensities. The origin of this phenomenon is thought to be the existence of two distinct populations of functional units which are tentatively identified with the inner and outer hair cells. Thus $N_1(I)$ might belong to radial fibres from the inner cells and $N_1(II)$ to spiral innervations of the outer cells. The maximum amplitude of the first component is approximately ten times that of the second, suggesting a tenfold difference in the two populations. This is in agreement with the anatomical observations. With increasing stimulus, the high-intensity component grows much more rapidly than the lower one (figure 4.13).

At a given frequency, the latency of the response is a decreasing function of the stimulus level and again there is a difference in the behaviour of the two components. This is seen most clearly if the amplitude of the response is plotted against its latency, as illustrated in figure 4.14. The function shown in this diagram was obtained by a

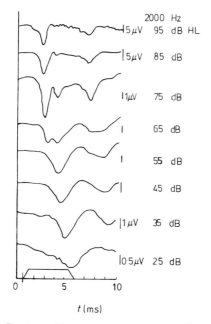

Figure 4.12   Electrocochleograms for a person with normal hearing. The stimulus is a 2 kHz tone burst (see text). The envelope of the stimulus is indicated at the bottom of the diagram and its onset coincides with the appearance of the cochlear microphonic. The tracings are shown for senation levels in the range 25–95 dB. From Eggermont *et al* (1974).

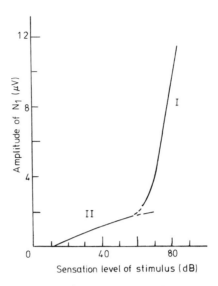

Figure 4.13 Amplitude of the N_1 response in a person with normal hearing as a function of the stimulus level. The stimulus is a 2 kHz tone burst (see text). The two intersecting functions relate to the two components of the response as indicated (cf figure 4.12). Redrawn (simplified) from Eggermont *et al* (1974).

least-squares analysis of data from normal human subjects. The amplitudes and latencies so represented are those of the larger component, whichever this happens to be. Thus for an individual subject the function would be discontinuous at intermediate intensities where there is a transition from one component to the other.

As indicated in figure 4.11 the cochlear action potential is poorly developed at frequencies below 1500 Hz. This is also a feature of the brainstem response described in §4.6. The reason for the loss of signal at low frequencies is partly that the response is inherently weaker at these frequencies and partly that its latency is relatively less stable. Variations in latency (known as 'jitter') impair the synchronisation of successive responses and broaden the waveform of the average response which then becomes difficult to detect.

4.6 Evoked Vertex Potentials

Electrodes on the scalp provide a far-field indication of activity in the brain. The potential changes they register are due primarily to coordinated post-synaptic activity in the dendrites which make up the bulk of the grey matter of the cerebral cortex. Nevertheless, well synchronised discharges in some nerve trunks can also produce a measurable response. Electrical activity in the brain is continuous throughout life and is detected outside the skull as a fluctuating potential in which certain rhythmic patterns can be recognised. A record of this activity is called an electroencephalogram. Its abbreviation (EEG) is frequently used to denote spontaneous, as distinct from evoked, activity recorded by a far-field electrode, usually on the scalp. The evoked potentials are generally at least an order of magnitude smaller than those associated with spontaneous events, but unlike the latter they have a definite temporal relationship to the stimulus and so can be detected by averaging.

In the last 15 years an enormous interest has developed in the

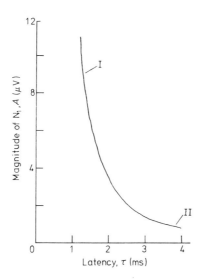

Figure 4.14 Amplitude, A (μV) of the N_1 response as a function of its latency, τ (ms) in the electrocochleogram of a person with normal hearing. The stimulus is a 2 kHz tone burst (see text). The function is represented by the equation $\log A = 1 \cdot 21 = 2 \cdot 22 \log \tau$, which was obtained by a least-squares analysis of the data for six subjects. The values of A and τ relate to whichever component of N_1 happens to be the greater. For each subject the function is discontinuous at the transition from one component to the other. Data and equation from Eggermont *et al* (1974).

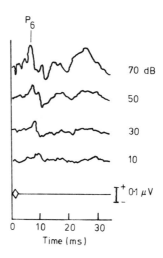

Figure 4.15 Brainstem response for a person with normal hearing. The potential is recorded at the vertex with a reference electrode on the earlobe. The stimulus is a tone pip with a central frequency of 2 kHz as indicated at the bottom of the diagram. The tracings shown are averages of 2048 responses with an inter-stimulus interval of 40 ms. The numbers indicate the sensation level of the stimulus. Redrawn from Davis 1976 *Ann. Oto-Rhino-Laryngol.* **85** suppl. 28.

measurement of neurogenic response to auditory stimulation. This is demonstrated in a review by Reneau and Hnatiow (1975) which cites nearly 500 works on the subject, most of which were published between 1960 and 1970. It is probably the availability of small averaging computers that has made possible the expansion of work in this field over so short a period. The motive for this endeavour is the hope that electric response audiometry will solve the problem of diagnosis in infants and other difficult patients. At a fundamental level the study of auditory evoked potentials is also a study of brain function, and as such it is a subject beyond the scope of this book. There would be little merit in presenting a catalogue of the various vertex potentials but in order to give a more complete picture of the electric response the basic features of two of these potentials will be described. The reader wishing to learn more about the auditory evoked potentials and their role in audiometry will find the monograph by Davis (1976) a useful starting point. As with cochlear potentials, these responses are usually elicited by clicks or tone bursts. The active electrode is at the top of the head; its exact position is not critical.

4.6.1   Fast vertex potentials

These potentials have their origin in events in the auditory nerve and the brain stem (see table 4.1). A number of positive maxima with latencies ranging from 2 to 12 ms have been identified (Jewett and Williston 1971). The first of these is probably generated in the auditory nerve but the remainder come from the brain itself and provide the so-called brainstem electric response (BSER). By far the most prominent feature of this response is Jewett's Vth wave which has a characteristic latency of about 6 ms and is designated P_6 (see figure 4.15). The amplitude of P_6 is extremely small (less than $0 \cdot 2 \, \mu V$) but fortunately the response tolerates a high stimulus repetition rate so that as many as 2000 samples can be collected and averaged in a few minutes. With good equipment the response remains detectable down to $0 \cdot 03 \, \mu V$, which corresponds to a stimulus level near to the threshold of hearing.

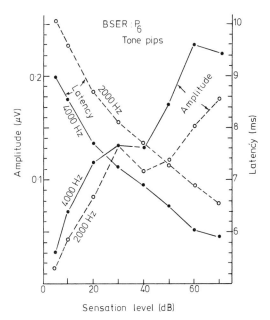

Figure 4.16   Amplitude and latency of the P_6 response shown in figure 4.15. From Davis 1976 *Ann. Oto-Rhino-Laryngol.* **85** suppl. 28.

The latency is closely related to that of the cochlear potential N_1 (figure 4.16).

4.6.2 Slow vertex potential

The slow vertex potential designated N_{90}–P_{180}–N_{250} in table 4.1 is particularly interesting because it exists in virtually the same form for several different sensory modes (figure 4.17). For this reason it is sometimes called a non-specific response, but there is evidence to show that its origin is chiefly the primary cortical projection area for the modality concerned (for audition this is a region in the superior surface of the temporal lobe beneath the sylvian fissure). In common with most evoked potentials the slow vertex response is a transient event. It is particularly sensitive to *changes* in the stimulus such as an abrupt change in frequency, intensity or interaural time difference.

The amplitude of the slow response is typically about $15\ \mu V$ and thus is much greater than the brainstem potential. When evoked by tone bursts or clicks the magnitude and latency have a distinctly

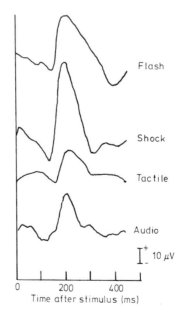

Figure 4.17 Slow vertex response (nominally N_{90}–P_{180}–N_{250}) for comfortably strong' stimuli of various forms. Redrawn (simplified) from Davis 1976 *Ann. Oto-Rhino-Laryngol.* **85** suppl. 28.

nonlinear relationship to the level of the stimulus. The response becomes detectable close to the threshold of hearing and increases rapidly with stimulus intensity up to a sensation level of 20 dB. Thereafter it grows only slowly until at 80 dB the amplitude is approximately double that at 20 dB. A further increase in intensity produces unpredictable results. The latency is constant for stimuli up to 30 dB, but below this level it increases as the stimulus is reduced, and near threshold it has approximately double its normal value.

The brainstem response (P_6) and the cochlear response (N_1) are favourites for diagnostic audiometry because they are reliable and stable phenomena capable of providing specific information about the auditory system. The cochlear response is of course a monaural reaction, but the brainstem response may be obtained from either ear and is slightly increased when both are stimulated simultaneously. Electrocochleography requires anaesthesia and the services of a surgeon, whereas brainstem audiometry merely involves the use of a sedative to prevent interference from muscle potentials. The slow cortical response is also valuable for audiometry providing the subject is fully awake and cooperative and is not restless or fatigued. Unlike cochlear and brainstem responses, the slow vertex potential is practically independent of frequency throughout the audible range and in suitable subjects it can provide reliable estimates of the auditory threshold at any desired frequency.

5 Hearing Disorders and the Measurement of Hearing

5.1 Introduction

This chapter is concerned mainly with the role of audiometry in the diagnosis and evaluation of hearing disorders. Information about the performance and calibration of audiometric equipment will also be given. Before describing the diagnostic procedures it is necessary to present a brief account of the causes of deafness. Hearing impairments due to such agents as disease, injury and congenital abnormality will be described, but the etiology and treatment of these disorders will not be considered. For further information the text by Mawson (1974) is recommended.

5.2 Disorders of Hearing

Most, if not all, auditory defects are associated with a loss of sensation though this is not necessarily the only complaint. The loss of sensation can be expressed in terms of a corresponding elevation of the threshold of hearing, and the term *hearing loss* (HL) is then used to denote the difference in decibels between the threshold of a defective ear and the threshold of a normal ear exposed to the same stimulus. A similar term, namely *hearing level*, is also used but it has more than one meaning: when applied to the subject it means hearing loss as just defined, but when applied to the stimulus it means *sensation level* (SL). The latter was defined in Chapter 1 as the intensity level of a sound relative to its intensity at threshold. In this context it should be noted that the reference threshold, 0 dB (SL), is not necessarily that of a normal ear.

Hearing loss (deafness) is called *conductive* if it is the result of a defect in the conductive mechanism; *sensorineural* (neuro-sensory) if its origin is in the cochlea or auditory nervous system; and *non-organic* (psychogenic) if it has no apparent bodily cause.

5.2.1 Conductive deafness

There are several causes of conductive deafness. These are, with the exception of an obstruction of the external ear, defects of the middle ear. In principle it is also possible that the inner ear might have a conductive disorder which would impair the transmission of mechanical energy to an otherwise normal organ of Corti and it has been suggested (Davis and Silverman 1970) that this type of lesion may contribute to the degeneration of hearing with advancing age (presbycusis). There is, however, no direct evidence for its existence.

A common cause of conductive deafness is damage to the tympanic membrane and ossicles as a consequence of infection. Varying degrees of damage are encountered from simple perforation of the eardrum to the almost complete destruction of the ossicular system.

In children the most common conductive disorder is the occurrence of a middle ear effusion. In this condition the tympanic cavity is filled with fluid apparently secreted by the mucosal lining of the middle ear. The consistency of the fluid varies enormously from a watery 'serous' liquid at one extreme to a thick visco-elastic 'glue' at the other. The associated hearing loss is, however, independent of the physical nature of the effusion (Bluestone *et al* 1973). Middle ear effusions are always accompanied by a blockage of the Eustachian tube and inadequate ventilation of the middle ear; indeed, it is generally believed that Eustachian tube dysfunction is a precursor of the disease. If, as a result of tubal dysfunction, the tympanic cavity is effectively sealed, absorption of oxygen by the middle ear tissues reduces the intratympanic pressure (Rasmussen 1967, Elner 1977) so that the eardrum is drawn inwards. This in itself produces a hearing loss, though the impairment is mild (10–15 dB) and is confined to low frequencies.

Another disease which produces a conductive hearing loss is otosclerosis. This is a hereditary disease in which parts of the otic capsule are invaded by softer bone which grows sporadically and then hardens. A common site for the new growth is close to the antro-inferior margin of the oval window. In about one case in ten the growth spreads to the stapes and as the latter becomes fixed (ankylosed) in the oval window, a progressive hearing loss develops. The late stages of the disease are accompanied by a progressive sensorineural loss in addition to the conductive impairment just described.

Conductive hearing loss is generally distributed fairly uniformly in the frequency range normally considered for audiometry (250–4000 Hz), although the impairment is often emphasised at low frequencies. It is unusual to find a high-tone conductive loss. The general features of conductive loss in relation to the disorders described above are as follows.

(*a*) Perforation of the tympanic membrane creates a low-frequency bypass of the eardrum impedance, but unless the damage is extensive the high tones are unaffected.

(*b*) Middle ear scarring, fibrosis and adhesions which are often the sequelae of infection reduce the efficiency of the middle ear transformer. At low frequencies the middle ear impedance is stiffness-controlled and therefore not well matched to the resistive impedances of the meatus and cochlea (§2.3). Thus an increase in stiffness due to disease is accompanied by an increase in the transmission loss for low-frequency sound. Mild disorders, such as reduced intratympanic pressure, which increase the stiffness of the eardrum or ossicular lever, have their effect mainly at low frequencies but the loss spreads to higher frequencies with increasing severity of the damage to the middle ear and in many cases the picture is one of a fairly uniform loss at all frequencies.

(*c*) The presence of an effusion in the middle ear reduces the compliance of the tympanic cavity simply by displacing the air which would normally occupy this space and the mastoid air cells. It also increases the effective mass of the eardrum. The corresponding hearing loss ranges from about 15 to 50 dB and affects all frequencies, although it is often greatest below 1 kHz.

(*d*) In otosclerosis the disease affects only the 'output end' of the middle ear transformer, leaving the remainder of the conductive mechanism undamaged. In the early stages the hearing loss is greatest in the middle and low audiometric frequencies. There is also a curious loss of sensitivity for bone-conducted sound in the neighbourhood of 2 kHz, which is known as the Carhart notch. In the later stages of the disease the hearing loss becomes fairly uniform at all frequencies.

Conductive deafness is seldom, if ever, total. This is because high-intensity sounds transmitted through the skull stimulate the inner ear directly. The sensitivity of the cochlea to bone-conducted sound depends on the difference in acoustic impedance between the round and oval windows, and in a defective ear this depends on the nature and extent of the middle ear damage. As a general guide it can be reckoned that the complete absence of the middle ear mechanism results in a hearing loss of 50–60 dB. Because the auditory system has a large dynamic range its performance is usually thought of in terms of a logarithmic scale. For this reason it often appears that the ear can tolerate a fairly extensive modification of its conductive apparatus. Thus a disorder which produced a tenfold reduction in the amplitude of the stapedial motion would create a 20 dB hearing loss. This would generally be regarded as a mild disability. Unlike sensorineural lesions, conductive defects merely attenuate the incoming acoustic signal without introducing any significant distortion. The intelligibility of the signal can therefore be restored, if necessary, by amplification.

5.2.2 Sensorineural hearing loss

The term sensorineural implies a lesion in either the organ of Corti (sensory) or some unspecified part of the auditory nervous system (neural). The term is deliberately non-specific so that it applies to all forms of deafness which are not conductive or psychogenic. It is a description of a symptom rather than its cause, for it is practically impossible to identify the site of the lesion on the basis of hearing loss alone. At best, audiometric tests can sometimes distinguish between two classes of lesion, namely, those occurring in the cochlea itself, and those occurring on the VIII nerve or brainstem (retrocochlear lesions). The appropriate tests will be described later but it should be noted that retrocochlear disease is rare and that in the great majority of cases sensorineural hearing loss can be presumed to have a cochlear origin. Sensorineural impairment is undoubtedly the most prevalent form of hearing disorder for there is scarcely a person past his youth in whom its presence cannot be demonstrated. Thus whatever hearing defects may additionally occur, there is an underlying trend for the performance of the inner ear to diminish with advancing age. This deterioration, known as presbycusis, is the most common cause of deafness in the elderly. There are, however, many other causes of sensorineural hearing loss including congenital defects of the cochlea, injury, ototoxic agents, and local or systemic infections. The loss of function can be anywhere from a mild high-tone loss to complete deafness. Usually, but not always, the loss is greatest for the high frequencies and in some cases it can increase dramatically with frequency, possibly at a rate of 50 dB per octave. This incidentally demonstrates the tuning capabilities of even a defective ear.

Sensorineural deafness is usually accompanied by distortion and a reduced ability to discriminate speech in the presence of noise. The intelligibility of speech is not completely restored by amplification as it would be in the case of a conductive loss. Another feature of sensorineural impairment is a compression of the dynamic range of the ear, known as recruitment. Thus weak or moderate sounds may not be heard at all while intense sounds may evoke a normal or near-normal sensation of loudness. With some neural lesions recruitment is diminished or even absent. This will be considered later in connection with the diagnosis of retrocochlear disease.

5.2.3 Tinnitus

Many hearing disorders, including otosclerosis, are accompanied by tinnitus. This is a condition in which the subject has the sensation of hearing sounds which have no external origin. The noises heard are usually high-pitched tones or bands of noise but they may also be bizarre low-frequency rumblings and 'chugging' sensations. Tinnitus may be present in one or both ears. Its subjective loudness (as determined by loudness balance measurements) is close to threshold and the noises are readily masked by an external sound if this is audible. Despite its apparently trivial loudness tinnitus can be a distressing condition, particularly if the patient's threshold of hearing is so high that ambient sounds fail to mask the subjective noise.

Tinnitus almost certainly has more than one cause. Often it seems to originate in the cochlea since it occurs in association with disorders of the inner ear. On the other hand, it can be experienced by patients who have had the auditory nerve divided at surgery. In some cases the noises are pulsatile, demonstrating a connection with the vascular system.

5.2.4 Menière's disease

This disease of the inner ear is named after Prosper Menière who first described it. The chief symptoms are a triad consisting of vertigo, deafness and tinnitus. Unlike other disorders of the inner ear, Menière's disease is episodic and between attacks there are often long periods in which the sufferer is free of symptoms. It is generally believed that the disease is a failure of the mechanism regulating the production or disposal of endolymph, resulting in an over-production of endolymph (endolymphatic hydrops) and a gross distension of the scala media and saccule.

5.2.5   Tumours affecting the VIII nerve

The most common of these tumours is an acoustic neuroma. The tumour expands slowly within the internal auditory meatus, compressing the nerve and the blood vessels serving the inner ear, so that in its early stages it gives rise to both neural and labyrinthine symptoms affecting one ear. If untreated, its enlargement causes severe brain damage and ultimately death. Other cerebello-pontine angle tumours can also affect the VIII nerve and produce similar clinical features.

5.2.6   Psychogenic hearing loss

A loss of hearing or an inability to understand speech can sometimes have a central cause even though an organic lesion cannot be identified. This problem is partly in the province of the neurologist or psychiatrist and will not be considered here. As far as audiometry is concerned the task is to identify true non-organic deafness and to distinguish this from malingering. The problem is complicated by the fact that both organic and psychogenic disorders can occur simultaneously.

5.3   Threshold Audiometry

The most important of all audiometric procedures is the measurement of auditory thresholds for pure tones. With few exceptions the stimulus is delivered through earphones (air conduction) or by means of a mechanical vibrator applied to the skull (bone conduction). The threshold to be determined is the level of the stimulus at which the probability of a positive response is approximately 0·5, although as explained in §3.2, it is not necessary to adhere strictly to this definition. In practice the threshold is the least audible sound estimated to the nearest 5 dB and variations due to the use of different psychoacoustic techniques are generally neglected. The usual procedure is to present the tone at a constant level for one or two seconds and to follow this with a silent interval. Alternatively, the tone may be interrupted automatically about two or three times per second during its presentation. The subject indicates whether or not he heard the stimulus and its intensity is then adjusted accordingly.

Audiometers are calibrated directly in decibels relative to a standard representing the threshold of normal hearing. Compensation for the variation of the reference sound pressure level with frequency is made

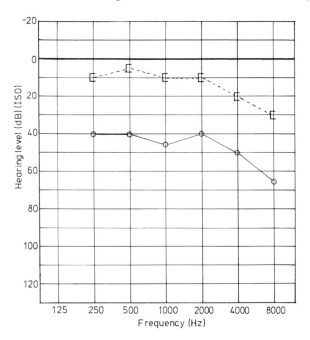

Figure 5.1 Audiogram showing a hearing loss in the right ear. The deafness is conductive with a sensorineural component at high frequencies. The standard symbols used on audiograms are: air conduction × left, ○ right; bone conduction] left, [right. A recommended format for audiograms is given in *J. Laryngol. Otol.* **89** 1069–74 (1975).

automatically. By common agreement thresholds are always plotted in the same way using a form of diagram called an audiogram. This representation is illustrated in figure 5.1, which shows a hypothetical right-sided hearing loss. The standard audiometric frequencies are plotted from left to right and hearing loss (hearing threshold level) is given on the ordinate. The horizontal line at 0 dB represents normal hearing. Points above this line indicate hearing that is better than the standard; points below indicate impairment. The total hearing deficit is shown by the air conduction curve and the sensorineural component by the bone conduction curve. The difference between these curves is known as the 'air–bone gap,' and this represents the conductive component of the hearing loss.

5.3.1 Reliability of threshold measurements

The air conduction thresholds shown in figure 3.3 are distributed as shown in figure 5.2. The dispersion of the thresholds is the result of intersubject variability and it needs to be considered when assessing individual audiograms. The standard deviations of the thresholds about the mean value depend on frequency, but are generally about 6 dB. This rather small dispersion relates only to the performance of young, well-motivated people tested under ideal conditions. Considerably greater dispersion is likely in the hurly-burly of clinical audiometry. Thus an apparent hearing loss of 20 dB or less at any one frequency is

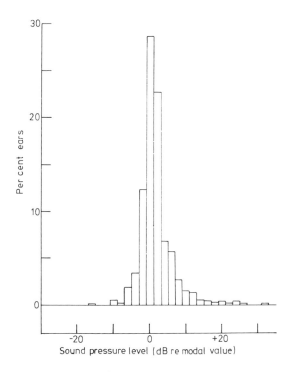

Figure 5.2 Distribution of the sound pressure level at the threshold of hearing at 4096 Hz. Similar distributions are obtained at other frequencies. The histogram shows the results for 1028 ears (514 subjects) considered to be otologically normal. Note that the distribution is asymmetrical so that more people have thresholds above the modal value than below it. Redrawn (with minor modifications) from Wheeler and Dickson 1952 *J. Laryngol. Otol.* **66** 379–95.

probably not significant, but a loss of 15 dB or more at several frequencies probably does indicate an abnormality. The chief source of error in air conduction measurements with earphones is the variability in acoustic impedance of the ear due to individual differences in the volume of the meatus. This is discussed further in §5.9. The error is present at all frequencies. An additional difficulty at high frequencies is that the sound pressure is not uniform throughout the meatus and important changes in the pressure distribution may accompany small changes in the position of the earphone. At low frequencies the pressure is uniform and the position of the earphone is not critical, except that good acoustic seal should be formed with the pinna. Acoustic leak is a major source of error at frequencies from 500 Hz downwards.

Statistics for normal bone conduction thresholds show that these have a similar dispersion to the air conduction thresholds (Whittle 1965). Sources of error in bone conduction measurements are as follows.

1. There are large anatomical variations in the size of the skull.

2. Bone vibration transducers are difficult to calibrate and reliable standards have yet to be developed (see §5.10).

3. The stimulatory effect of a bone vibrator is sensitive to changes in the position of the vibrator and the force used to apply it to the surface of the skull.

The bone vibrator is usually placed on the mastoid on the same side of the head as the ear being tested. Forehead placement can also be used although most audiometers are not calibrated for this position. Békésy (1960) has stated that the position of the vibrator is less critical if forehead as opposed to mastoid placement is used, and this has been confirmed by Hart and Naunton (1961). The preference for siting the vibrator on the mastoid is therefore curious but a possible explanation is that it is difficult to deliver an intense undistorted stimulus through bone conduction. The slightly greater sensitivity at the mastoid (table 5.6) can be used to advantage to extend the working range of the audiometer.

5.4 Masking

In threshold audiometry each ear is tested separately so that it is often necessary to mask one ear in order to prevent it receiving an audible stimulus while the other ear is being tested. The interaural attenuation of a stimulus delivered through earphones is approximately 50 dB, whereas negligible attenuation exists for a bone-conducted sound even when the vibrator is on the mastoid. Thus air conduction measurements can be made without masking provided that the difference in hearing levels of the ears is less than 50 dB, but bone conduction measurements generally require masking.

The best form of masking is a narrow band of noise having a bandwidth approximately equal to the critical band at the appropriate frequency (see §3.7). This type of noise has the minimum loudness for a given degree of masking, a feature which is important because excessive noise is uncomfortable and distracting for the subject, and in an extreme case can produce an acoustic reflex in the ear being tested.

There are several recipes for determining the correct level of masking but the technique described by Hood (1957) is certainly the best. In this method the sensation level of the masking sound is increased in steps of 10 dB from 0 dB upwards and the patient's threshold determined at each level of masking. If masking is in fact necessary, the threshold determined in this way will increase with the level of masking up to the point where the signal is heard in the ear being tested. Thereafter no further increase in threshold occurs unless the masking noise is made so intense that it is detected in the contralateral ear (figure 5.3).

As will now be explained, the maximum degree of masking that can be used depends on the amount of vibration imparted to the skull by the source of the masking noise, and on the bone conduction threshold of the ear under investigation. An earphone is not only a source of air-conducted sound but also a source, albeit a weak one, of bone-conducted energy. The bone-conducted component is, in terms of sensation level for normal hearing, at least 50 dB less than the air-conducted component. Thus for practical purposes an interaural attenuation of 50 dB can be assumed if earphone masking is used. This can be increased to 80 or 90 dB by using as the source of noise, a hearing aid receiver fitted with a soft rubber tip which is inserted into the meatus (Littler *et al* 1952). There are, however, two disadvantages to the use of an insert. The first is that its output cannot be calibrated. The second and more serious disadvantage is that unless special care is taken in fitting the insert, it can easily be displaced during the test and so cause an unintended and unnoticed change in the level of masking. Earphones and insert receivers also radiate airborn sound which is

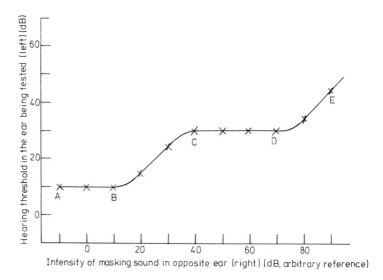

Figure 5.3 Illustration of Hood's method of masking. In this example the left ear is being tested by bone conduction and the right ear is being masked. A–B, the masking sound is inaudible and the tone is heard in the right ear; B–C, the masking sound is audible but the tone is still heard in the right ear; C–D, the tone is now heard only in the left ear; D–E, the masking sound is now sufficiently intense that despite interaural attenuation its component in the left ear also has a masking effect. The subject's true threshold in the left ear is 30 dB.

received in the contralateral ear. This mode of interference is unimportant because the associated interaural attenuation is greater than that for the direct transmission of energy from an earphone (or insert) to the skull.

5.5 Rainville Audiometry

A method of bone conduction audiometry which does not involve the contralateral ear has been described by Rainville (1955) and Jerger and Tillman (1960). In this technique, wide-band noise is delivered through a bone vibrator while at the same time a pure tone is presented through an earphone. The level of the noise that just masks the pure tone at threshold intensity is determined. The masking effect of the noise is known from a previous calibration and hence the bone conduction threshold can be estimated. It should be noted that this threshold is determined with the ear occluded by the earphone. Thresholds measured in this condition are generally lower than those for an unoccluded ear because small changes in the volume of the meatus are produced by the mechanical disturbance of the skull and create perceptible pressure changes when the meatus is restricted. For a discussion of this effect see the articles on bone conduction by Békésy (1960) and Naunton (1963).

5.6 Automatic Audiometry (Békésy)

The audiometer to be described was invented by von Békésy (1947) and its use is often called Békésy audiometry. The instrument delivers a pure tone, heard through earphones, and is automated in such a way that the subject plots his own threshold audiogram. The frequency and intensity controls of the audiometer are driven by electric motors. During the test the frequency increases at a slow but steady rate. The intensity also changes at a pre-determined rate but the direction of the change (increase or decrease) is controlled by the subject whose task is then to maintain the intensity close to his threshold of hearing. The intensity is increased automatically until the tone becomes audible and at this point the subject closes a switch which causes the intensity to decrease. As soon as the sound becomes inaudible he releases the switch and the process is reversed. The audiogram is drawn on a printed chart by a pen, the movement of which is derived from the setting of the frequency and intensity controls and the tracing so obtained has a saw-tooth pattern. In an alternative version of the instrument (McMurray and Rudmose 1956), discrete rather than continuous changes in frequency are produced so that each of the standard audiometric frequencies are presented in turn. In this case the horizontal movement of the pen is a function of time rather than frequency, but the chart is divided into a series of intervals, each corresponding to a particular frequency, so that the overall pattern is similar to that of a conventional audiogram.

The principal application of Békésy audiometry is in industry, where it provides a rapid method of screening workers whose jobs involve exposure to potentially hazardous levels of noise. It also has, perhaps fortuitously, a diagnostic role in the detection of retrocochlear disease (Jerger 1960).

5.7 Loudness Balance: Recruitment

As already observed, recruitment is an abnormal growth of loudness
with intensity of the stimulus. The phenomenon can best be demonstrated by what is known as the alternate binaural loudness balance
(ABLB), a technique invented by Edmund Fowler (1936). In this test the
tone is presented to each ear in turn. The intensity at one ear (the
reference ear) is kept constant while the intensity at the other ear is
adjusted so that equal sensations of loudness are obtained. A number
of loudness matches are determined for different values of the reference intensity. Fowler devised this procedure to test his hypothesis
concerning otosclerosis. He thought that in the early stages of the
disease the amplitude of the stapedial motion would be limited to
some critical value determined by the narrowing of the annular ligament, but that smaller movements would be unrestricted. Thus loudness should increase with intensity until the limiting amplitude is
reached and thereafter remain constant. Fowler called this the 'hobble
effect.' Eventually the hypothesis had to be abandoned for neither
hobble nor recruitment could be demonstrated in ears with 'clinically
established uncomplicated otosclerosis.' But Fowler had invented a test
which has remained in regular use for over 40 years.

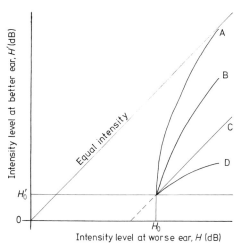

Figure 5.4 Possible results of the binaural loudness balance test
(ABLB). For explanation see text.

Figure 5.4 is a classification of the results that can be obtained in a loudness balance test. Each curve is a plot of the intensity at one ear against the intensity at the other for the balance condition. Thus each line represents equal sensation. The intensities are referred to the threshold for normal hearing so that H'_0 and H_0 are the hearing losses in the better and worse ears, respectively. For a person with normal hearing the balance points lie on or close to the equal intensity line. The line labelled C running parallel to this is obtained if there is no recruitment, a result which occurs in conductive (and sometimes neural) deafness. Curve A is said to show complete recruitment if it meets the equal intensity line or over-recruitment if it crosses it. Curve B is an example of incomplete recruitment and curve D shows loudness reversal. Curve A indicates a cochlear lesion, whereas curves C and D are associated with neural disorders. The interpretation of B is equivocal. For further information on the interpretation of loudness balance tests see Priede and Coles (1974).

5.8   Speech Audiometry

Speech audiometry, as the name implies, uses voiced sounds rather than pure tones to convey the stimulus. The response is judged on the number of linguistic units that are correctly identified. Audiometry is the measurement of hearing, so that speech audiometry is properly regarded as a test of a person's ability to hear speech. The audiometric procedures can, however, be extended to include tests of the speech transmission system itself rather than the listener's auditory performance. For example, a telephone system could be assessed on the basis of speech tests and for this it would be appropriate to have listeners with normal hearing. The testing of hearing aids is another example, but in this case the listeners would have hearing impairments.

In diagnostic audiometry the spoken material is usually recorded on a gramophone record or magnetic tape and played to the listener through a speech audiometer which provides various degrees of amplification. The presentation is either 'free-field' or through earphones. Earphones are more convenient and have the advantage that one or other ear can be tested independently. Masking with wide-band noise can be used if required (Coles and Priede 1975). Free-field listening is necessary if the audiometry is designed to test the patient's performance with a hearing aid. In the United Kingdom the most widely

Table 5.1 Recorded speech audiometry materials in the UK. C = consonant, V = vowel. From Lyregaard et al (1976) with minor alterations.

Name of material	Type of material	References	No. of items per list	No. of lists	Recording medium	Word spacing (seconds)	Speaker/ accent	Made by/ obtainable at	Comments
MRC word lists	Phonetically balanced mono-syllables	Knight and Littler (1953)	25 mono-syllables	20				Royal National Institute for the Deaf, London	A new recording has been made at the RNID but this is not yet available
Fry word lists	Phonetically balanced mono-syllables	Fry (1961)	30 CVCs + 5 CV/VCs = 100 phonemes	10	Magnetic tape	4	Male voice/ Southern English	Dept of Phonetics University College, London	
Fry sentence lists	Phonetically balanced sentences	Fry (1961)	25 sentences (100 key words)	10	Magnetic tape	—	Male voice/ Southern English	Dept of Phonetics, University College, London	

Boothroyd word lists	Iso-phonemic mono-syllables	Booth-royd (1968)	10 CVCs (30 phonemes)	15	Magnetic tape	5	Male voice/ Northern English	Dept of Audiology and Education of the Deaf, University of Manchester	Each list contains exactly the same 30 different phonemes
					Magnetic tape	5	Male and Female BBC speaker	Royal National Institute for the Deaf, London	Recording includes two training words per list, plus one-minute running speech
					Magnetic tape	4	Male voice/ Southern English	ISVR, Southampton University	Lists 9, 10 and 15 not included in this recording

used speech material consists of monosyllabic nouns grouped into phonetically balanced lists, or lists containing at least the more common phonemes†. These lists are known as MRC, Fry and Boothroyd after their originators. Details are given in table 5.1.

One of the problems in speech audiometry is to specify the level of the stimulus. The term *speech level* is used to denote the average sound pressure level relative to 20 μPa measured at the ear. The sound pressure itself is easy to obtain; the difficulty lies in choosing a suitable method of averaging and a number of alternatives have been suggested (Lyregaard *et al* 1976). Two of these are as follows.

1. The speech signal is regarded as an amplitude-modulated carrier which may be demodulated by rectification or squaring followed by low pass filtering. The resulting signal is then integrated (averaged) in an RC circuit having a time constant which is short compared with the duration of a representative sample of speech material (suggested time constants range from 30 to 200 ms). The speech level is then defined with reference to a selected percentile in the distribution of the level of the integrated signal that is obtained from a sample of the speech material in question. An A-weighted filter may be incorporated to correct for the frequency dependence of the auditory threshold.

2. Speech level is defined with reference to the A-weighted L_{eq} of the original speech signal. It is then a measure of the acoustic energy associated with a sample of the speech material. For the definition of L_{eq} see Burns and Robinson (1970).

Fortunately, the problem of defining a physical reference level for speech does not occur in most diagnostic audiometry; instead the following procedure is used. Speech audiograms are obtained for normal subjects. Each audiogram is a plot of intelligibility against intensity level, where intelligibility (discrimination) is the fraction of words correctly identified, and intensity level is an arbitrary decibel value indicated on the output attenuator of the speech audiometer. The intensity corresponding to some arbitrary point on the normal intelligibility curve is then used as a reference and the output of the audiometer so referred is called the *relative speech level*. The usual reference is the intensity which gives a discrimination of 50%. An

---

† *Phoneme:* an abstract element of a morpheme, cf. a letter as an element of a word. *Morpheme:* the smallest linguistic element which has a meaning, cf. a word. *Phonetic balance* (PB) in a word list: occurrence of phonemes in the list is approximately that in 'everyday' speech.

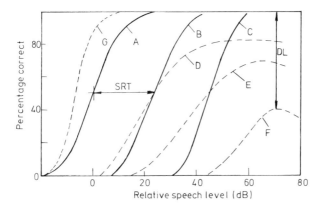

Figure 5.5   Speech audiograms: A and G, normal hearing; B and C, conductive deafness; D, E and F, sensorineural deafness; curve G is for sentences and the others are for phonetically balanced words. The relative speech level is referred to the intensity at which 50% of the words are heard correctly. Conversational speech lies at about 45 dB on this scale. Cases B and D have the same speech reception threshold (SRT) of 23 dB. The discrimination loss (DL) in case F is 62%. From Robinson (1971).

alternative suggested by Hood and Poole (1977) is the speech detection threshold, defined as the level at which 50% of the words in a given presentation can be heard but not necessarily understood.

Figure 5.5 shows typical speech audiograms for normal and impaired ears. For conductive loss the intelligibility curve is merely shifted to a higher level without a significant change in shape, but for sensorineural loss discrimination at first increases with intensity and then decreases after passing through a maximum which is less than 100%. Two useful parameters may be obtained from a speech audiogram. These are the *speech reception threshold* (SRT), defined as the relative speech level at which 50% of the words (or other scored items) are correctly identified, and the *discrimination loss* (DL) defined as 100 minus the maximum discrimination score. The above definitions apply to European usage (Robinson 1971). American terminology is somewhat different in that discrimination loss is measured not at the peak of the intelligibility curve but at a level 30 or 40 dB above the speech reception threshold (ANSI 1960). Thus the American literature sometimes describes cases in which the speech reception threshold is measurable but the discrimination loss is total.

5.8.1 Accuracy of speech audiometry

The determination of the intelligibility curve is a statistical procedure in which the discrimination at each intensity is obtained from the fraction c/N of scored items correctly identified by the subject in N presentations of the scored material. In order to estimate the accuracy of each point on the discrimination curve it is assumed that the probability p of identifying a scored item is the same for all N items presented. The best estimate of p is then c/N. If somehow the speech test could be repeated *ad infinitum*, the number of identified words in each test would be distributed binomially about the mean value Np with a standard deviation $\sqrt{Np(1-p)}$. If N is sufficiently large (say 100) and if p does not have an extreme value (e.g. $0 \cdot 1 < p < 0 \cdot 9$), then the binomial distribution is approximated by a normal distribution having the same parameters. It is then legitimate to express the discrimination score as $100(p \pm \delta p)\%$, where δp is the standard error in the estimate of p, namely, $\sqrt{p(1-p)/N}$. When the normal approximation does not apply there is no simple method of expressing the accuracy of the result, but upper and lower confidence limits may be stated. Values of the confidence limits for binomial distributions are tabulated in *Documenta Geigy* (Diem and Lentner 1970) and are displayed graphically in figures published in the *Biometrika* tables (Pearson and Hartley 1966).

The errors considered above are a consequence of the statistical nature of speech audiometry. They are therefore the least possible errors that would occur under ideal test conditions. Although other errors are usually small by comparison they are sometimes worth considering. In order to obtain a speech audiogram it is necessary to select a different list at each intensity so that the material is heard and not merely remembered from an earlier presentation. The words in a given list vary in their intrinsic intelligibility, that is at a given intensity some words are more frequently understood than others. This variability is considered desirable because it spreads the intensity range over which the intelligibility curve occurs. If all words were equally difficult the curve would rise abruptly with increasing intensity and its shape would then be relatively difficult to determine. It is of course important that all the lists should be as far as possible identical with regard to the distribution of difficulty in the words they contain. This property of the MRC material has been investigated by Hood and Poole (1977) who showed that most of the lists were very satisfactory in this respect, but that some improvement could be obtained if a few 'rogue' lists were excluded. They showed that further improvements could be made if each list were assigned its own reference for determining the relative speech level. Each list was tested with a group of normal listeners and a mean intelligibility curve obtained. The effect of choosing slightly different reference levels for each list was to bring these curves into coincidence. The work described by Hood and Poole was an evaluation of a new set of tape recordings of the MRC lists produced by the RNID†. These lists are not at present available but the intention is to make copies of the RNID tapes and in the process to incorporate the adjustments in level.

5.9   Audiometric Standards: Air Conduction Thresholds

As already explained, the threshold of hearing is a somewhat variable quantity. There are variations from one individual to another and, even for the same subject, different methods of testing may give different results. In order to compare the findings of different laboratories or clinics it is necessary to have a universal, if arbitrary, standard of normal hearing. This is particularly important if the hearing loss needs to be known for legal reasons such as the assessment of handicap and claims for compensation.

There are several problems in establishing standards for pure tone threshold audiometry. The first is to choose a suitable physical quantity to represent the magnitude of the stimulus. The quantity concerned must be one that can easily be measured or compared with other standards. Although such quantities as the displacement of the stapes or the sound pressure at the tympanic membrane might be considered ideal representations of the stimulus, they are difficult to measure and therefore not appropriate to a definition of auditory performance. It is considerably easier to measure sound pressure outside the ear and audiometric standards therefore relate to acoustic pressure in the unobstructed field (free-field audiometry) or at the entrance to the auditory meatus (earphone audiometry). At audio frequencies the determination of sound pressure in a free field is a relatively uncomplicated acoustic measurement and reference has already been made in Chapter 3 to the British Standard (BS 3383) defining the minimum audible field. Free-field audiometry with pure tones is rarely done except as a laboratory procedure. This contrasts with the widespread

† Royal National Institute for the Deaf, 105 Gower Street, London.

use of earphone audiometry and thus the majority of work has been the determination of international standards for earphone listening.

The standardisation of earphone thresholds has had a long and troubled history which can be traced through the bibliographies given in the various published standards. Difficulties arise both in the choice of suitable subjects and in the acoustic measurements themselves. The former British Standard (BS 2497: 1954) was based on the work of Dadson and King (1952) at the National Physical Laboratory, and Wheeler and Dickson (1952) at the Royal Air Force Central Medical Establishment. In this work the subjects were between 18 and 25 years old and special care was taken to exclude any that might have had hearing disorders. Subjects with low intelligence who may have been unable to cooperate fully with the test procedures were also excluded. This rigorous selection contrasts with earlier work (Steinberg *et al* 1940) which formed the basis of a now defunct American Standard†.

The measurement of sound pressure at the meatus involves the use of a small probe microphone connected to a tube inserted in the cap of the earphone. Such microphones are relatively insensitive so that the measurement of meatal sound pressure requires laboratory facilities. To overcome this difficulty, pressure measurements on real ears have been transferred to standard artificial ears or couplers which are then used in defining audiometric zero (see table 5.2).

5.9.1 Artificial ears and couplers

A coupler is a device for providing an acoustic connection between an earphone and a microphone. It usually consists of a cylindrical cavity of specified dimensions with the earphone at one end and a pressure microphone at the other. An artificial ear is a similar device except that it is designed so as to present the earphone with the same acoustic impedance as that of an 'average' human ear. Thus the microphone, if suitably placed, will register a sound pressure equivalent to that at the entrance to the human ear. The merit of an artificial ear is that it can, in principle, be used with any earphone so that a reference sound pressure corresponding to audiometric zero can be specified independently of the type of earphone. The only theoretical objection to this

---

† American Standard AZ 24.5 (1951). A comparison of this and the present standard is given in ANSI S3.6 (1969) Appendix D.

Table 5.2 Normal thresholds for monaural hearing with the STC 4206A earphone.

	rms sound pressure (dB re 20 μ Pa)			
	British Standard (1954)		Revised British Standard (1972) and ISO Standard	
Frequency (Hz)	1 Meatus	2 BS 2042 Artificial ear	3 BS 2042 Artificial ear	4 IEC Artificial ear
80	44·0	60·5	—	—
125	30·0	44·5	47·0	45·0
250	19·0	29·5	28·0	27·5
500	12·0	12·0	11·5	13·5
1000	9·0	6·0	5·5	8·0
1500	10·5	7·5	6·5	7·5
2000	11·0	9·0	9·0	10·5
3000	7·5	6·0	8·0	11·5
4000	9·5	9·0	9·5	13·5
6000	14·0	9·0	8·0	13·5
8000	18·5	9·0	10·0	16·0
10000	19·5	15·5	—	—
12000	22·5	—	—	—
15000	36·5	—	—	—

Notes

(*a*) All data in the table are modal values for otologically normal subjects aged 18–30 years. The thresholds apply equally to men and women (Dadson and King 1952).

(*b*) The psycho-acoustic technique is not specified in the standards. Dadson and King presented the tone several times, each presentation lasting approximately 2 s. After determining an approximate threshold, the final value was approached from above and below and the threshold taken to be the 'mean of the two lowest levels at which the tone was consistently heard by the subject.'

(*c*) The entries in columns 1 and 2 are taken from BS 2497 (1954). These data also appear in Dadson and King (1952). The artificial ear is described in BS 2042 (1953).

(*d*) The entries in column 3 are from BS 2497 part 1 (1968). This information also appears in ISO 389 (1975).

Notes (*contd.*)

> (*e*) The entries in column 4 are from BS 2497 part 3 (1972). This is a
> revision of BS 2497 (1954) and is technically identical to ISO 389. The
> artificial ear, described in BS 4669 (1971), conforms to IEC 318 (1970).
> The data in column 4 are applicable to a variety of earphones including the
> TDH 39—for details see BS 2497. The earphone is applied with a force of
> 5 N.

form of standardisation is that it does not allow for variations in
impedance among individual human ears. Since the output impedance
of an earphone is high compared with the impedance at the meatus, a
given excitation of the earphone will result in a slightly greater sound
pressure in an ear with a small meatus than in an ear with a large one.
At present there is no practical solution to this problem. According to
Robinson (1971) the uncertainty in the stimulus due to differences
between human ears is typically 2 or 3 dB but may sometimes be much
greater. Nevertheless, the standard deviations in Dadson and
Wheeler's measurements were not significantly different from those
relating to the free-field pressure measurements (Sivian and White
1933). But for the highest accuracy sound pressures have to be
measured with a probe microphone under the earphone.

Examples of the construction of artificial ears and couplers are
shown in figures 5.6 and 5.7. Not all artificial ears and couplers achieve
the ideal of an accurate representation of the real ear throughout the
range of audiometric frequencies, but instruments constructed to the
IEC Standard (IEC 318: 1970) are very satisfactory (Delany *et al*
1967).

5.9.2   Routine calibration of audiometer earphones

An artificial ear is not essential for routine calibration of earphones
since adequate results can be obtained with a simple coupler. When a
coupler is used, the reference sound pressures apply only to specified
earphone/coupler combinations. Most audiometers now in use in the
United Kingdom are fitted with the Telephonics TDH 39 earphones,
though some have TDH 49 or Telex 1470 earphones. These earphones
may be calibrated to the ISO standard on an NBS 9A coupler (or
Brüel and Kjaer type 4152, 6 cc artificial ear) using the reference levels
given in table 5.3.

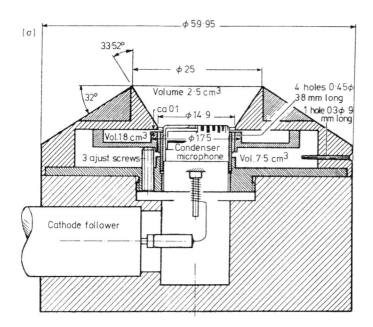

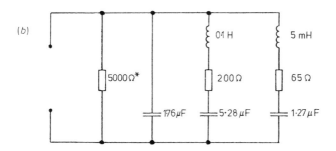

Figure 5.6 (a) An artificial ear constructed to the IEC specification. This specification is equivalent to BS 4669 (1971). A commercially produced version of this artificial ear is the Brüel and Kjaer type 4153. All dimensions in mm. (b) Equivalent circuit of the artificial ear. 1 electrical ohm corresponds to $10^5 \, \text{N s m}^{-5}$. The $5000 \, \Omega$ is the minimum permitted resistance of a pressure equalisation leak which may be coupled to any one of the three cavities. The $65 \, \Omega$ resistance is obtained by adjustment of the three screws shown in (a). From IEC 318 (1970). Reproduced by kind permission of the International Electrotechnical Commission, to which the copyright belongs.

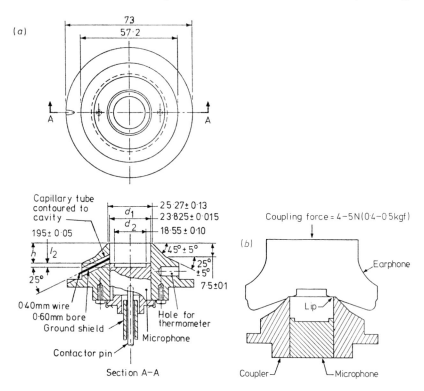

Figure 5.7 (a) An acoustic coupler (IEC reference type) based on the NBS 9A coupler and available commercially as the Brüel and Kjaer type 4152. The value of h may be varied to suit the equivalent volume of the microphone. In the 9A coupler, $h + l_2 = 13.41$ mm for all microphones. (b) Coupling of earphone to coupler. All dimensions in mm. From BS 4668 (1971), reproduced by permission of BSI, from whom complete copies can be obtained.

5.10 Audiometric Standards: Bone Conduction Thresholds

When held against the mastoid or forehead, the bone vibrator applies a small alternating force to the skull. The auditory stimulus can be expressed in terms of the rms value of this force or the corresponding acceleration of the skull at the point of application of the vibrator. The standards for normal hearing by bone conduction relate to the performance of the bone vibrator on an artificial mastoid designed to simulate the mechanical impedance of the human skull at the mastoid process.

Table 5.3   Reference equivalent threshold sound pressure levels in the NBS 9A coupler: ISO standard.

Frequency (Hz)	dB re 20 μPa		
	1 TDH 39	2 TDH 49	3 Telephonics 1470
125	45·0	47·5	45·0
250	25·5	26·5	25·0
500	11·5	13·5	10·0
1000	7·0	7·5	3·0
1500	6·5	7·5	5·0
2000	9·0	11·0	4·0
3000	10·0	9·5	5·0
4000	9·5	10·5	6·0
6000	15·5	13·5	7·5
8000	13·0	13·0	9·0

Notes

(*a*) Entries in column 1 are from ISO 389 (1975) and are identical to BS 2497 part 2 (1969) and ANSI S3.6 (1969).

(*b*) Entries in column 2 are from BS 2497 part 2 (1969).

(*c*) Entries in column 3 are from Causey and Beck (1974).

(*d*) The data for the TDH earphones apply to models in either the original metal shell or the new plastic one. Details are obtainable from Telephonics, 770 Park Avenue, Huntington, New York 11743. See also Michael and Bienvenue (1977).

(*e*) The meatal sound pressures corresponding to the figures in the table are not given in the ISO or British Standards. These publications merely state that 'the reference levels all refer, as closely as can be ascertained from existing data, to the same auditory threshold levels.' These levels can be inferred from the information given in table 5.2 by adding to the figures in column 1 the difference between the corresponding figures in columns 3 and 2. It will be seen that for most purposes this difference is of a negligible order.

(*f*) The earphone should in each case be fitted with an MX 41/AR cushion and applied to the coupler without acoustic leakage with a force of 4–5 N, excluding the weight of the earphone itself.

(g) The numbers in the table are sound pressures as determined by a pressure microphone in the coupler. When applying this information it should be remembered that a correction may be necessary for the frequency response of the microphone and measuring equipment. The output of the earphone is obtained by subtracting the reference levels in the table from the sound pressure level indicated on the measuring equipment.

Table 5.4 Mechanical impedance of an ideal artificial mastoid (IEC standard).

Frequency (Hz)	Mechanical reactance (N s m⁻¹)	Mechanical resistance (N s m⁻¹)	Mechanical impedance (N s m⁻¹)
125	−290·0	74	299·3
160	−220·0	55	226·8
200	−180·0	44	185·3
250	−140·0	36	144·6
315	−110·0	29	113·8
400	−89·0	25	92·4
500	−71.0	22	74·3
630	−55·0	20	58·5
800	−42·0	19	46·1
1000	−32·0	18	36·7
1250	−23·0	17	28·6
1500	−17·0	17	24·0
1600	−15·0	17	22·7
2000	−8·4	17	19·0
2500	−2·2	18	18·1
3000	+2·7	18	18·2
3150	+3·9	18	18·4
4000	+10·0	19	21·5
5000	+17·0	21	27·0
6000	+22·0	23	31·8

Notes

(*a*) The data in this table come from ANSI S3.13 (1972). This information appears identically in IEC R 373 (1971) and BS 4009 (1975), but in the latter two publications it is restricted to the frequency range 250–4000 Hz.

(*b*) The impedance is presented to a bone vibrator having a plane circular contact area of 175 mm² and applied with a static force of 5·4 N.

(*c*) The specified impedance is approximated by a three-component network having in series a mass of $0·77 \times 10^{-3}$ kg, a spring of stiffness $2·25 \times 10^{5}$ N m⁻¹ and a resistance of 19·3 N s m⁻¹ (ANSI S3.13: 1972).

The mechanical impedance of an ideal artificial mastoid is given in table 5.4. This specification is closely approached by the Brüel and Kjaer type 4390 artificial mastoid, the performance of which is shown in figure 5.8. The construction of the B and K mastoid (figure 5.9) is similar to that of an instrument built at the NPL and described in BS 4009 (1975). The resistive component and the negative part of the

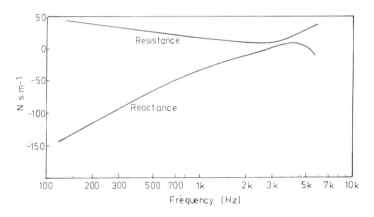

Figure 5.8 Mechanical impedance of a typical Brüel and Kjaer type 4930 artificial mastoid. Compare IEC specification given in table 5.4. Reproduced from Brüel and Kjaer Instruction Manual for Artificial Mastoid Type 4930.

mechanical reactance are provided by viscoelastic rubber pads between the vibrator and a loading mass which supplies the positive reactance. The output is obtained from a piezoelectric force transducer mounted on top of the inertia terminal ('seismic mass') as shown in the diagram.

An alternative arrangement—the Beltone model 5—described by Weiss (1960) is mentioned in the ANSI standard and has been used in

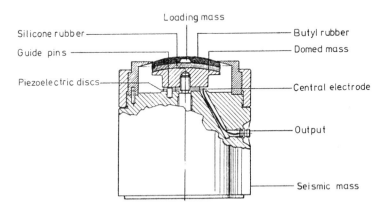

Figure 5.9 Construction of the Brüel and Kjaer type 4930 artificial mastoid. Reproduced from Brüel and Kjaer Instruction Manual for Artificial Mastoid Type 4930.

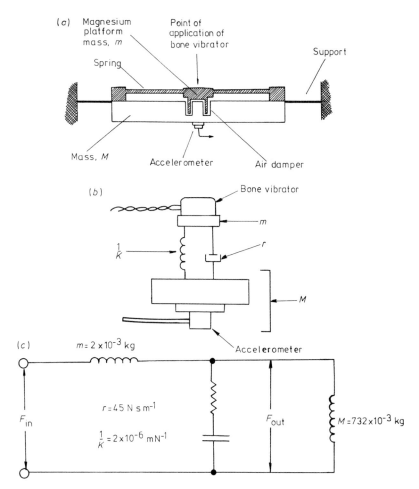

Figure 5.10 (*a*) Simplified drawing of the Weiss air-damped artificial mastoid on which the Beltone model 5 is based. The bone vibrator is applied to a light magnesium platform suspended on three springs (also of magnesium). The mass of the platform, including the effective mass of the springs, is 2 g. The suspended system is mounted on a larger mass M (732 g) to which is attached an accelerometer. Relative movement of the two masses is damped by a precision air damper—the annular gap is only 0.05 mm. From Weiss (1960). Reproduced with permission of the American Institute of Physics. (*b*) and (*c*) Schematic representation and equivalent circuit of the artificial mastoid shown in (*a*). Reproduced with permission from ANSI S3.13, © 1972 by the American National Standards Institute, copies of which may be purchased from the ANSI at 1430 Broadway, New York 10018, USA.

the development of bone vibrators manufactured in the USA. The essential features of this device are shown in figure 5.10. The bone vibrator is held against a light platform of mass m which is coupled to a much larger mass M by means of resistive and reactive elements. The damping is provided by the viscosity of air contained in a narrow annulus between the two masses. The ratio of the force on the small mass to that on the large mass is approximately constant. The exact relationship is given in table 5.5 which also shows the input impedance calculated from the equivalent circuit. It will be seen that this impedance does not meet the IEC requirements and in this respect the NPL/B and K mastoid is superior. Both forms of artificial mastoid have to be calibrated by means of an electromechanical transducer which generates a known acceleration at the input.

At present (1979) reference thresholds for bone conduction are not prescribed in the International Standard although recommendations are made elsewhere. The British Standard (BS 2497 part 4: 1972), recognising the errors inherent in bone-conduction measurements, specifies a reference threshold acceleration of $-30\,\mathrm{dB}$ re $1\,\mathrm{m\,s^{-2}}$ at all frequencies in the range 250–6000 Hz (see table 5.7). This acceleration is to be measured on the artificial mastoid specified in BS 4009, that is, the NPL or Brüel and Kjaer type. The standard requires the bone

Table 5.5 Data for a typical Beltone model 5 artificial mastoid.

Frequency (Hz)	Force ratio (dB)	Mechanical reactance (N s m^{-1})	Mechanical resistance (N s m^{-1})	Mechanical impedance (N s m^{-1})
250	−0·1	−432·0	86·0	441·0
500	−0·3	−164·0	52·0	172·0
750	−0·7	−99·0	48·0	110·0
1000	−1·1	−68·0	46·5	82·5
1500	−1·8	−34·5	46·0	57·0
2000	−2·0	−15·0	45·5	48·0
3000	−1·0	+11·5	45·0	46·5
4000	+0·9	+30·5	45·0	54·5

Notes

(a) The information in this table is taken from ANSI S3.13 (1972), Appendix A. The impedances are calculated for the equivalent circuit shown in figure 5.10.

(b) The force ratio is 20 log (force on m/force on M).

vibrator to have a plane circular contact area of $175\,mm^2$ and to be applied with a force of 5·4 N. Audiometers for use in the United Kingdom are currently calibrated to this standard. Since in many cases the bone vibrator is contoured to adapt to the shape of the human mastoid, the above restriction concerning the contact face is generally ignored.

The American Standard (ANSI S3.13: 1972) provides interim reference thresholds, expressed in terms of the rms value of the alternating force exerted by the vibrator on the artificial mastoid (table 5.6). Reference accelerations derived from the force levels are also given in the American Standard, but as these values are calculated with reference to the mechanical impedance of the Beltone artificial mastoid

Table 5.6   Reference equivalent threshold force levels for calibration of bone vibrators (dB re 1 μN).

Frequency (Hz)	1 Mastoid	2 Forehead	3 Mastoid
250	63·0	76·5	61·5
500	57·5	72·5	50·5
750	49·0	61·5	39·5
1000	43·0	53·0	37·0
1500	40·5	49·5	35·5
2000	40·0	48·5	28·0
3000	30·5	38·0	26·5
4000	35·0	41·5	31·0

Notes

(*a*) Entries in columns 1 and 2 are from ANSI S3.13 (1972). These are interim values and apply to audiometers with an air conduction output calibrated to ANSI S3.6: 1969 (equivalent to ISO 389).

(*b*) The interim values are intended to apply to vibrators calibrated on the Beltone artificial mastoid, although this is not stated unambiguously in the American Standard. The corresponding force levels on the Brüel and Kjaer mastoid are given in column 3. The data in this column are from Wilber (1972) expressed to the nearest 0·5 dB.

(*c*) All values in the table are for monaural listening with the test ear unoccluded.

(*d*) The vibrator is to be held against the skull with a static force of 3·9 N.

Table 5.7 Reference equivalent threshold acceleration levels for calibration of bone vibrators.

	dB re 1 m s^{-2}		
Frequency (Hz)	1 ANSI (Beltone)	2 BS 2497	3 Oslo
250	−46·0	−30·0	−34·0
500	−37·5	−30·0	−30·0
750	−38·5	−30·0	−31·0
1000	−39·5	−30·0	−33·0
1500	−35·0	−30·0	−35·0
2000	−31·5	−30·0	−30·0
3000	−37·5	−30·0	−28·0
4000	−32·0	−30·0	−26·0

Notes

(*a*) Data in column 1 are from ANSI S3.13 (1972) and apply to the Beltone model 5 artificial mastoid.

(*b*) Data in column 3 are from the Brüel and Kjaer instruction book for the type 4930 artificial mastoid and are recommendations based on work at the Institute of Audiology, Oslo. These levels are said to be under consideration for a future IEC standard.

(*c*) The British and American Standards specify that the thresholds are for monaural listening with the test ear unoccluded. The vibrator is to be applied to the mastoid process with a force of 5.4 N (BS) or 3·9 N (ANSI).

(*d*) Acceleration levels for forehead placement can be obtained by comparison with the data in table 5.6.

(table 5.5), they are not directly comparable with the British Standard. The ANSI recommended acceleration levels are given in table 5.7.

Table 5.6 includes reference force levels for the Brüel and Kjaer mastoid. These values were determined by Wilber (1972) and provide a direct comparison between force levels on the Beltone and B and K mastoids for a given excitation of the bone vibrator. The data were obtained using a Radioear type B 70 A vibrator which does not conform to the IEC specification, but similar results would be expected for the B 71 vibrator which does meet the standard. A possible error in Wilber's comparison is that a weakness has been reported in the performance of the B and K mastoid used in her experiments (S F Lybarger 1978 personal communication). This is likely to have influenced the results in the neighbourhood of 2 kHz. The problem with the B and K mastoid was an inconsistency in the resistive component of the impedance, but it is understood that this difficulty has been overcome with the development of new rubber pads.

5.11   Performance Specifications for Audiometers

Specifications for the performance of audiometers are given in BS 2980 (1958) and ANSI S3.6 (1969). The following recommendations, based unless otherwise stated on the British Standard, are given here for the benefit of anyone engaged in testing audiometers.

1. The frequency of the signal should be within 3% of the indicated value for an audiometer generating discrete frequencies. There is no corresponding recommendation in the British Standard when the frequency is continuously variable, except a statement that as far as possible audiometers having a continuously variable frequency should perform as well as discrete-frequency equipment. The American Standard specifies a tolerance of 5% when the frequency is continuously variable.

2. The level of any harmonic should be at least 30 dB below the level of the fundamental. Audiometer earphones, such as the TDH 39, are inherently free from distortion even at outputs in excess of 100 dB. Any significant distortion that occurs is usually present also in the voltage waveform at the terminals of the earphone. Bone vibrators†, on the other hand, do suffer from distortion particularly at frequencies from 250 Hz downwards. This often limits the useful output at these frequencies.

3. Operation of the tone interrupter switch must not generate audible transients. When the switch is turned on, the time for the output voltage to rise to 90% of its final value should not exceed 500 ms. The time for the voltage to rise from 10 to 90% of the final value should not be less than 50 ms. When the switch is turned off the time for the output voltage to fall to 0·1% of its original value should not exceed 500 ms. The decay from 90 to 10% of the original value

---

† Many audiometers are fitted with Radioear B 70 A or B 71 vibrators. Performance specifications, including information on distortion, can be obtained from Radioear Corporation, 375 Valley Brook Road, Canonsburg, Pennsylvania 15317.

should occur in not less than 50 ms. Operation of the switch should not at any time cause the output voltage to exceed the steady value by more than 1 dB.

4. With the tone interrupter switch in the off position the output should be 80 dB below its value when the switch is on, or 20 dB below the normal threshold of hearing, whichever is the greater.

5. It is important that the left and right channels of the audiometer function independently so that the signal or masking noise delivered through one earphone is not heard in the opposite earphone. The British Standard does not make any specific recommendation about this, but the American Standard requires that any signal in an earphone on the ear not being tested should be 70 dB below the signal in the ear under test, or 10 dB below the normal threshold of hearing, whichever is the greater.

6. The accuracy of the output attenuator should be ±1 dB for a nominal step of 5 dB. A proportionately smaller tolerance is required if the interval between attenuator settings is less than 5 dB. The accumulated error over any number of steps should not exceed ±3 dB. The sound pressure level measured in an artificial ear or coupler should be within 5 dB of the nominal output level for all settings of the attenuator and frequency control. This performance is adequate for audiometers used for 'screening' but for diagnostic purposes a slightly higher accuracy (say ±3 dB) would usually be expected.

5.12 Noise Levels in the Test Environment

The maximum permissible noise level in the audiometric test room is given in ANSI S3.1 (1960). The recommendations in this standard are given variously as octave, third octave and spectrum levels of the noise that can be tolerated without masking of the signal at the normal threshold of hearing. The standard applies only to air-conduction audiometry with the specified earphones and cushions (for example, TDH 39 and MX41/AR). Robinson (1971) has criticised this Standard for its leniency. Far lower levels of ambient noise would normally be expected in a good environment for diagnostic audiometry. The ultimate test of good sound-proofing is that it should allow the measurement of thresholds by bone conduction in a person with excellent hearing. The determination of such thresholds is therefore an appropriate empirical method for the evaluation of the test environment.

6 Acoustic Impedance and the Measurement of Middle Ear Function

6.1 Introduction

The measurement of acoustic impedance at the eardrum can provide useful information about the state of the middle ear. Impedance meters suitable for clinical use are described in this chapter and the diagnostic applications of this form of auditory testing are considered.

6.2 The Measurement of Acoustic Impedance

The impedance to be measured is the acoustic impedance at the eardrum, namely Z_r/S^2, where S is the area and Z_r the radiation impedance of the eardrum (see equation (1.21)). The measuring instruments always employ a probe which is inserted into the external meatus and thus the impedance to be measured is always 'seen' through this probe and the residual volume of the ear canal between the tip of the probe and the tympanic membrane. At frequencies below 1 kHz the probe and the meatus can be regarded as discrete components in the acoustic circuit, but at higher frequencies they need to be treated as transmission systems with distributed resistance and reactance. The measurement of eardrum impedance at frequencies above 1 kHz is a laboratory procedure requiring relatively complicated equipment (see for example Møller 1960), whereas low-frequency measurements can be made with simpler equipment suitable for routine clinical use. High-frequency impedance measurements do not yet have a diagnostic application but this may be a subject for future developments (Colletti 1975, 1976).

Much of the progress in the medical application of acoustic impedance measurement is due to the pioneering work of Metz. In his investigations Metz (1946) used an acoustic bridge based on a design due to Schuster (1934). This form of bridge was later refined by Zwislocki (1963) and produced commercially by the Grason–Stadler Company. The Schuster–Zwislocki bridge was directly analogous to one of the alternating current forms of the Wheatstone bridge used in the determination of electrical impedance. One arm of the bridge contained the acoustic impedance of the ear and this was balanced against a variable acoustic impedance having resistive and reactive components that could be adjusted independently. The balance point was determined by listening for the minimum sound in an earpiece connected between opposite arms of the bridge. The Zwislocki bridge was eventually ousted by an electroacoustic instrument designed by Terkildsen and Nielsen (1960) in which the only acoustic components, other than the ear itself, were a microphone and a sound generator. In this instrument the microphone generated an output voltage proportional to the sound pressure in the ear. The phase and amplitude of this voltage, relative to the voltage at the terminals of the acoustic source, were determined by balancing the microphone output against a reference voltage which could be adjusted for phase and amplitude, and these adjustments provided an indication of the acoustic impedance at the probe. The instrument was calibrated using hard-walled cavities which had a purely reactive impedance, but it was not possible to obtain a satisfactory calibration in terms of acoustic resistance, and attempts to use the equipment for resistance measurements were abandoned. Terkildsen and Nielsen called their apparatus an electroacoustic bridge, claiming that it was equivalent to Zwislocki's arrangement. It is difficult to see why this should be so because it is not possible to represent the electroacoustic instrument in terms of an equivalent four-branch network. The word 'bridge' has remained in the literature, although applied to electroacoustic impedance meters it is a misnomer because these instruments are not bridges in the sense understood by electrical engineers.

For diagnostic purposes an impedance meter has to perform two functions: to measure the acoustic impedance of the ear, and to record the changes in this impedance that occur when the static pressure in the external ear is varied. The latter procedure, known as tympanometry, will be described later but it is mentioned here to explain why the impedance of the ear will be regarded as a variable quantity.

At frequencies below 1 kHz the impedance of the meatus is effectively in parallel with the impedance at the eardrum even though these components are apparently in series. The reason for the parallel arrangement is that the wavelength of a low-frequency sound is large compared with the length of the meatus so that the sound pressure is uniform throughout the ear canal and equal to the pressure at the eardrum. The flux at the entrance to the meatus is the sum of the volume displacement at the eardrum and the flow which creates the acoustic condensation of the air in the meatus. The two impedances are therefore in parallel. The parallel configuration means that it is generally more convenient to work in terms of admittance rather than impedance. The admittance of the meatus, which is of no interest, can then be subtracted from the admittance at the probe to obtain the admittance at the eardrum.

The acoustic section of a typical impedance meter is shown in figure 6.1. An oscillator having a variable output voltage e supplies a transducer M_1. This generates a sound pressure p, proportional to e, at the entrance to a narrow tube T_1. The tube provides an acoustic coupling between the source and a probe inserted in the ear. A microphone M_2 is similarly connected through a tube T_2 and generates a voltage e' proportional to the sound pressure p' in the ear. For tympanometry a third tube T_3 is connected to a pump which controls

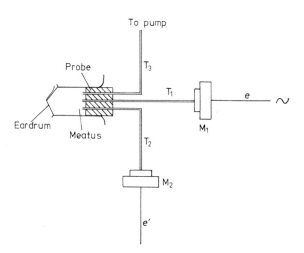

Figure 6.1 The acoustic components of an electroacoustic impedance meter. M_1 sound source; M_2 microphone; T_1, T_1 and T_3 connecting tubes; e voltage at input terminals of M_1; e' voltage generated by M_2. For further explanation see text.

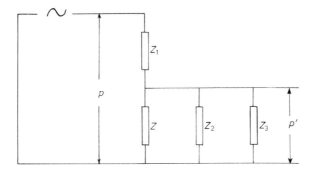

Figure 6.2   Equivalent circuit of the arrangement shown in figure 6.1.

the static pressure in the ear. The equivalent circuit of this arrangement is shown in figure 6.2. In this diagram the acoustic impedances of
the tubes T_1, T_2 and T_3 are represented by the discrete components
Z_1, Z_2 and Z_3, where Z_2 includes the impedance of the microphone
and the coupling between it and the tube. The impedance of the ear at
the exit of the probe is Z, and this is shunted by Z_2 and Z_3 so that the
combination has an impedance Z' given by

$$\frac{1}{Z'} = \frac{1}{Z} + \frac{1}{Z_2} + \frac{1}{Z_3} \tag{6.1}$$

or, in terms of admittance,

$$Y' = Y + Y_2 + Y_3 . \tag{6.2}$$

The microphone generates an output proportional to the sound pressure p' at the probe where

$$p' = Z'p/(Z_1 + Z'). \tag{6.3}$$

Thus

$$p/p' = 1 + Z_1/Z' = 1 + Z_1 Y' . \tag{6.4}$$

The ratio p/p' is a complex quantity which depends on the real and
imaginary components of Z_1 and Y'. If $|Z_1 Y'|$ is large compared with
unity, then

$$|Y'| = \frac{|p|}{|Z_1| . |p'|} . \tag{6.5}$$

Thus if the output of the oscillator is always adjusted so that the output of the microphone remains constant, the magnitude of the admittance at the probe is directly proportional to $|p|$ and hence to the voltage between the terminals of the source. The admittance at the probe is the sum of Y_2 and Y_3 and the admittances of the eardrum and the air in the meatus. In an ideal arrangement, Y_2 and Y_3 are made negligibly small so that Y' is approximately equal to Y. Furthermore, Z_1 is made large so that in combination with M_1 it constitutes a constant current source. In practice, the mechanical impedance of the sound generator may provide the requisite source impedance, and the impedance of the connecting tube may then be comparatively small. In this case, p and Z_1 should be regarded as quantities in an equivalent acoustic circuit representing the source. Similar considerations apply at the microphone. Both Z_1 and Z_2 need to be resistive so that the acoustic volume current is in phase with the oscillator voltage and so that the sound pressure p' is in phase with the output of the microphone. When these conditions apply, equation (6.4) gives

$$p/p' = R_1 Y = R_1(G + iB), \qquad (6.6)$$

where R_1 is the acoustic resistance of the source and G and B are the conductance and susceptance at the probe, respectively (for definitions see equation (1.27)). At low frequencies the meatus is a pure compliance C_m having a susceptance B_m given by

$$B_m = \omega C_m = \omega V/\gamma p_0, \qquad (6.7)$$

where ω is the angular frequency, V is the volume between the end of the probe and the eardrum, γ is the ratio of the principal specific heats of air and p_0 is the static pressure (see equation (1.28)). Thus if the subscript d denotes physical quantities at the eardrum, these are given by

$$p/p' = R_1[G_d + i(B_d + B_m)]. \qquad (6.8)$$

The ratio p/p' of the sound pressures is equal to the ratio e/e' of the source voltage to the microphone voltage and its real and imaginary components are independent measures of the acoustic conductance and susceptance. This principle is used in the Grason–Stadler model 1720 otoadmittance meter (figure 6.3). In this instrument the output of the microphone is used to control the electrical supply to the sound source so that a negative feedback loop is established which maintains a constant sound pressure in the ear. Thus e' is a constant while e

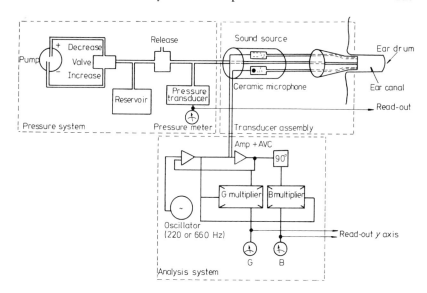

Figure 6.3   The Grason–Stadler model 1720 otoadmittance meter. The pressure system controls the static pressure in the external ear and is used for tympanometry. For explanation see text. Courtesy of Grason–Stadler, Inc.

varies in proportion to the admittance being measured. The phase and amplitude of the voltage at the source are determined by a comparison with the output voltage of the microphone, either directly or after the introduction of a 90° phase change. In simpler instruments (figure 6.4) the feedback is accomplished by a manual adjustment of the oscillator voltage. This adjustment brings the microphone output to a predetermined level as indicated by comparison with a reference voltage. Either the voltage at the source or the setting of the control which regulates it may be used to provide an indication of the magnitude of the acoustic admittance. Alternatively, the excitation of the source may be kept constant and the output of the microphone rectified and averaged to give a DC indication proportional to the sound pressure in the ear. In this mode of operation the output of the impedance meter is usually obtained from the difference between the reference voltage and the rectified microphone voltage. The indication is then a linear function of the magnitude of the acoustic impedance at the probe provided that Z_1 is large compared with Z'. Many impedance meters at present in clinical use are of the simpler variety in that they do not

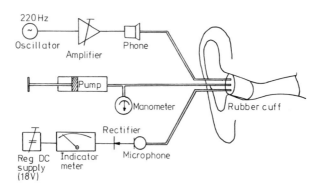

Figure 6.4 Electroacoustic impedance meter. This is a simplified version of the instrument shown in figure 6.3. For explanation see text. Courtesy of Grason–Stadler, Inc.

distinguish between resistive and reactive components of impedance. They provide an output proportional to $|Z|$ or $|Y|$ and, in the case of admittance, this is often displayed on a scale calibrated in equivalent volume units (ml) and is labelled 'compliance.'

Impedance meters usually operate at 220 or 275 Hz although higher frequencies, for example 660 Hz, are also available on some instruments. At 220 Hz the impedance at the probe is typically $10^8 \, \text{N s m}^{-5}$. Thus the resistive impedances Z_1, Z_2 and Z_3 should ideally be at least $10^9 \, \text{N s m}^{-5}$. An impedance of this order is easily achieved in the case of Z_1 and Z_2 because, as already mentioned, the transducers themselves rather than the connecting tubes can provide the necessary impedance at low frequencies. If miniature transducers are used these can be mounted directly in the probe so that the connecting tubes T_1 and T_2 may be only a few millimetres long and of a diameter which presents a negligible impedance. This arrangement provides the greatest efficiency because there is no unnecessary attenuation of the signal. The static pressure in the ear is controlled through the tube T_3 which consists of a flexible rubber pipe 1–2 m in length. This is usually connected to the probe by means of a capillary which provides the impedance Z_3.

The volume of the meatus between the end of the probe and the eardrum is of no clinical interest. Unfortunately this volume varies considerably from one ear to another, and unless some compensation is made for it the impedance at the probe is too variable a quantity to have any diagnostic significance. The volume of the meatus can be measured if the probe is made to terminate in a detachable speculum which can be left in the ear once the probe itself is removed. The volume between the end of the probe and the eardrum can then be determined by measuring the quantity of alcohol required to fill it. Although this method can provide an accurate measurement of meatal volume it is clearly not a convenient clinical procedure. The development of a simpler alternative followed the observation, attributed to Metz (1946), that small changes in the static pressure in the external ear produce large changes in impedance. The admittance of the eardrum and the middle ear structures is decreased when these are displaced from their normal position by the application of a pressure difference between the external and middle ear. If this pressure difference is made sufficiently large the admittance at the eardrum becomes negligible compared with the admittance of the air in the meatus. Under these conditions the admittance at the probe is proportional to the volume of the meatus and is numerically equal to it if the instrument is calibrated in equivalent volume units.

6.3 Tympanometry

Tympanometry is the measurement of acoustic impedance of the ear as a function of the static pressure in the auditory canal. A graph of this function is called a tympanogram. These definitions are not restricted to impedance, but also apply when admittance or compliance—or a related quantity such as sound pressure—is the measured variable. Figure 6.5 is a tympanogram for a normal ear. The quantity represented in this diagram is the combined admittance of the eardrum and the meatus. The admittance at the eardrum is a maximum when the static pressure difference across the tympanic membrane is zero. Hence the maximum in the tympanogram occurs when the pressure in the external ear is equal to the pressure in the middle ear. Thus the middle ear pressure may be determined from the location of the peak in the admittance function or the notch in the corresponding impedance function. The validity of this indirect method of measuring middle ear pressure has been confirmed by Eliachar and Northern (1974) although some discrepancies between middle ear pressures measured tympanometrically and the same pressures measured directly using a manometer have been reported (Renvall and Holmquist 1976).

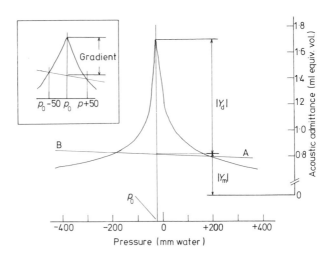

Figure 6.5 Admittance tympanogram at 275 Hz for a normal ear. The ordinate shows the magnitude of the acoustic admittance (or compliance) at the exit of the probe expressed in equivalent volume units. The abscissa shows the static pressure in the external ear relative to ambient atmospheric pressure. Middle ear pressure is p_0. The inset shows schematically, and on a different scale, the region around the peak of the tympanogram and illustrates the definition of gradient. For further explanation see text.

For a normal ear, admittance changes rapidly with pressure in the neighbourhood of the maximum, but it becomes relatively insensitive to variations in pressure once the difference across the tympanic membranes exceeds ±200 mm water. For practical purposes the admittance at the probe may be regarded as constant for pressures in the external ear greater than +200 mm water, and this admittance provides a basis for estimating the volume of the meatus. The straight line AB in figure 6.5 shows the admittance of a hard-walled cavity having the same volume as the estimated volume of the meatus. The acoustic admittance associated with this volume is numerically equal to the susceptance B_m and depends on pressure in accordance with equation (6.7). However, a pressure change of 100 mm water produces only a 1% change in this admittance so that the pressure dependence can usually be ignored.

The admittance at the probe is given by

$$|Y| = |Y_d + Y_m|, \tag{6.9}$$

where Y_m is the admittance of the air in the meatus and Y_d is the admittance at the eardrum. As already stated, the admittance of the meatus is that of a pure susceptance at the frequencies usually employed (220 Hz). For most applications Y_d is also treated as though it derives from a susceptance, so that the total admittance Y is simply the algebraic sum of Y_d and Y_m. This being the case, it is permissible to divide values on the ordinate in figure 6.5 by the angular frequency ω and so obtain compliance rather than admittance as a function of pressure. Admittance tympanograms are often presented as compliance tympanograms and the distinction is unimportant provided that the resistive component of the eardrum impedance is negligible. The resistive component can usually be disregarded at 220 Hz but not at 660 Hz (see table 6.3 and also figure 2.8).

6.3.1   Assessment of tympanograms

The shape of a tympanogram is often greatly modified by the presence of a conductive abnormality and, to some extent, the diagnosis of a conductive disorder can be made without recourse to quantitative measurement merely by noting the general pattern of the tympanogram (Jerger 1970). Qualitative assessment is satisfactory when the interpretation is obvious, but it is inadequate for cases which do not come into this category. For these, quantitative evaluation is often helpful. A further consideration is that the shape of a tympanogram depends on the physical quantity represented (admittance or impedance) and the linearity of the measuring system.

Tympanograms can provide the following basic information.

(a) *Middle ear pressure:* This is usually the static pressure in the external ear when the admittance is a maximum. With some middle ear disorders bimodal tympanograms are obtained if the probe frequency is above 220 Hz. However the maxima are not usually separated by a large change in pressure so that the average value between the peaks is a satisfactory indication of middle ear pressure.

(b) *Admittance at the eardrum:* This admittance is normally measured when the static force on the eardrum is zero, that is, at the peak of the tympanogram. The result is often expressed in terms of compliance, so that

$$C_d = C - C_m, \qquad (6.10)$$

where the subscripts have the same meanings as in equation (6.9). The compliance C_m of the meatus is usually obtained at a static pressure of 200 mm water above atmospheric pressure. If the eardrum compliance is small, it may be worthwhile correcting for the pressure dependence of C_m as discussed earlier. It may also be necessary to correct for the change in the response of the impedance meter itself to pressure. This additional pressure dependence is due to variations in the acoustic impedance of the connections to the source and the microphone and to variations in the sensitivity of the transducers themselves. The appropriate correction can be determined empirically by coupling the probe to a hard-walled cavity and varying the pressure.

The magnitude of the admittance peak in a tympanogram is influenced to a small degree by the rate at which the pressure in the external ear is changed. For normal ears the admittance increases with the speed at which the test is performed (Creten and van Camp 1974). This phenomenon has not been systematically investigated but a tentative explanation is that it is due to the viscoelasticity of the middle ear tissues. It may be that the eardrum becomes temporarily flaccid during a period of delayed elastic recovery from the deformation caused by the previous application of pressure. This is consistent with the observation that the admittance does not increase with the rate of change of pressure in ears which are already flaccid (Williams 1976 unpublished, see Feldman 1976). In order to avoid complications due to possible dynamic effects of the changing pressure, tympanograms should be performed with a slowly varying pressure. Alternatively, if the pump is operated manually, the rate of change of pressure can be brought to zero by pausing at the point where the peak admittance is to be measured. Compliance determined at constant static pressure is often called 'static' compliance (Jerger 1972). The term is unsatisfactory because a compliance determined acoustically is necessarily a dynamic quantity measured at the frequency of the probe tone. In other branches of physics, static compliance denotes the ratio of a constant displacement (or change in volume) to the constant force (or pressure) producing it.

(c) *Gradient:* A useful measure of the shape of a tympanogram is the slope of the curve on either side of the maximum. A suitable numerical expression of this quantity, called the 'gradient,' was devised by Brooks (1968, 1969), and is defined as the mean reduction in compliance for pressure changes in the external ear of ±50 mm water relative to the pressure at which the compliance is a maximum (figure 6.5). The quantity so defined is not a gradient in the strict mathematical sense, but it is difficult to find an alternative word to describe it. The chief merit of gradient as a diagnostic parameter is that it is independent of the volume of the meatus and the admittance associated with it. There is frequently some uncertainty about the value of this admittance because tympanograms often retain an appreciable slope even at high positive pressures, and the admittance presented to the probe at +200 mm water is sometimes a poor approximation to the true admittance of the meatus.

6.4   Acoustic Impedance Measurements as an Aid to the Diagnosis of Middle Ear Disorders

Before considering in detail the tympanometric identification of middle ear disease, some general remarks concerning the place of tympanometry in relation to other diagnostic procedures are appropriate.

A conductive abnormality usually gives rise to an abnormal tympanogram, although this is not invariably the case. Conversely, an abnormal tympanogram is not necessarily an indication of a significant conductive disorder. The starting point in the diagnosis of conductive deafness is therefore the demonstration of a significant difference between the air and bone conduction thresholds. Tympanometry can then be used to confirm the audiometric findings and perhaps to identify the lesion when this is not obvious from the patient's history or from an otoscopic examination. The value of tympanometry is greatly enhanced if it is coupled with otoscopy and, ideally, impedance testing should always be preceded by an examinations of the ear with a good quality otoscope. This will often reveal abnormalities such as a perforation or gross middle ear scarring which may have an important bearing on the interpretation of the tympanometric findings. Tympanometry is a much easier and quicker test to administer than threshold audiometry and it does not require the subject's active participation or a quiet environment. It is therefore ideal as part of a school screening procedure (Brooks 1973, 1974, 1976, Harker and van Wagoner 1974).

Table 6.1 Conversion of equivalent volume to basic CGS and MKS units.

	Admittance		
	CGS		MKS
Hz	$cm^4\,g^{-1}\,s$	mmhos	$m^4\,kg^{-1}\,s$
220	$0{\cdot}98\times10^{-3}$	0·98	$0{\cdot}98\times10^{-8}$
275	$1{\cdot}22\times10^{-3}$	1·22	$1{\cdot}22\times10^{-8}$
660	$2{\cdot}93\times10^{-3}$	2·93	$2{\cdot}93\times10^{-8}$
	Impedance		
	CGS		MKS
Hz	$cm^{-4}\,g\,s^{-1}$	ohms	$m^{-4}\,kg\,s^{-1}$
220	$1{\cdot}02\times10^{3}$	1020	$1{\cdot}02\times10^{8}$
275	$0{\cdot}82\times10^{3}$	820	$0{\cdot}82\times10^{8}$
660	$0{\cdot}34\times10^{3}$	340	$0{\cdot}34\times10^{8}$

Compliance in equivalent volume units can be converted to CGS or MKS units of acoustic admittance by multiplying the equivalent volume by $\omega/\gamma p_0$.

1 ml of air at standard atmospheric pressure $(1{\cdot}013\times10^5\,N\,m^{-2})$ has a compliance of $7{\cdot}05\times10^{-12}\,m^4\,kg^{-1}\,s^2$ $(7{\cdot}05\times10^{-7}\,cm^4\,g^{-1}\,s^2)$ and the admittance or impedance shown in the table.

6.4.1 Normal ears

The acoustic admittance at the eardrum of ears with normal middle ear function has been investigated by a number of workers including Brooks (1969, 1971) and Jerger *et al* (1972). The results of these surveys are given in table 6.2 where the data are presented, as originally reported, in terms of compliance expressed in equivalent volume units. To facilitate comparison with results given in basic units, a list of conversion factors is provided in table 6.1. Brooks' results are for children in the age group 4–11 years, whereas Jerger's subjects were mainly adults. In Brooks' survey, separate distributions of admittance were obtained for boys and girls and for right and left ears, but no significant differences were detected. On the other hand, Jerger and his associates found that the mean value of the admittance was slightly higher in men than in women. They also reported that the admittance decreased slightly with increasing age for both men and women over the age of 40, but in view of the large variance in their observations this result may not be statistically significant.

Feldman (1974) has measured acoustic susceptance and conductance in normal ears using a Grason–Stadler model 1720 otoadmittance meter (table 6.3). These measurements were made with the external ear at ambient atmospheric pressure. In normal ears, middle ear pressure seldom differs from atmospheric pressure by more than

Table 6.2 Distribution of compliance at the eardrum and gradient of the tympanogram for normal middle ears.

		0·5	2·5	10	Percentile 50	90	97·5	99·5
		Compliance in ml equivalent volume						
Jerger *et al*	*N*							
(1972)	1650	—	0·30	0·39	0·67	1·30	1·65	—
Brooks								
(1969, 1971)	1394	0·28	0·35	—	0·70	—	1·40	1·72
		Gradient in ml per 50 mm water						
Brooks								
(1969)	1394	0·06	0·10	—	0·30	—	1·00	1·50

Notes

(*a*) In both surveys the compliance at 220 Hz was measured at middle ear pressure using a Madsen ZO-70 impedance meter. This instrument, although calibrated in terms of compliance, responds to the modulus of the acoustic admittance. The values in the table can be converted to admittance or impedance in basic units by multiplying by the factors given in table 6.1.

(*b*) In Brooks' survey the compliance of the air in the meatus was taken to be the equivalent volume with a pressure in the external ear of +200 mm water relative to atmospheric pressure. Presumably Jerger used the same method though this is not explicitly stated in the original publication. No correction was made for the change in compliance of the meatus with pressure.

(*c*) *N* is the number of ears. In Jerger's survey the distribution is that of the compliance averaged for both ears of each individual. In the original publication *N* is given as 825 (subjects).

(*d*) Criteria for normal middle ear function.
Brooks: in all subjects acoustic reflexes were obtainable in both ears with a contralateral stimulus of 95 dB sL (frequency not specified).
Jerger: in all subjects acoustic reflexes were obtainable in both ears with contralateral stimuli at 500, 1000 and 2000 Hz (intensity not specified). All subjects were required to have middle ear pressures in both ears greater than −100 mm water. Subjects giving grossly abnormal tympanograms were rejected. Approximately half of Jerger's subjects had sensorineural hearing loss.

Table 6.3 Components of the acoustic admittance at the eardrum for normal middle ears (Feldman 1974).

Percentile	Susceptance B_d		Conductance G_d		Admittance $|Y_d|$	
	220 Hz	660 Hz	220 Hz	660 Hz	220 Hz	660 Hz
10	0·30	0·90	0·05	1·00	0·33 (0·34)	1·38 (0·47)
50	0·50	1·30	0·15	1·95	0·52 (0·53)	2·43 (0·83)
90	0·75	2·45	0·35	4·20	0·85 (0·87)	4·50 (1·54)

Notes

(*a*) The values in the table are in millimhos, except those in brackets which are equivalent volumes in ml. To convert to MKS units multiply by 10^{-8} (table 6.1).

(*b*) The measurements relate to 100 ears (children and adults). The criterion for normal middle ear function was either a hearing level better than 10 dB at all audiometric frequencies in the range 250–4000 Hz, or the existence of identical thresholds for air and bone conduction at these frequencies.

(*c*) The method of compensating for the admittance of the air in the meatus is not stated in Feldman's paper, but presumably the components of this admittance were determined at a pressure of +200 mm water in the external ear.

(*d*) The susceptance and conductance at the eardrum were determined at ambient atmospheric pressure in the external ear.

30 mm water (Holmquist 1976), but small pressure differences across the tympanic membrane produce significant changes in its acoustic admittance and this may explain why Feldman's values are lower than those reported by Brooks and Jerger.

6.4.2 Middle ear effusions

Tympanometry is the most reliable non-invasive method for detecting fluid in the middle ear. Middle ear effusions in children are a common occurrence and in recent years there has been an increasing interest in this disease. The widespread use of tympanometry is due largely to its success in the diagnosis of this condition.

Middle ear effusions occur in association with inadequate ventilation of the middle ear and reduced intratympanic pressure. As explained in §5.2, the presence of fluid in the middle ear reduces the acoustic admittance at the eardrum by displacing the air in the middle ear

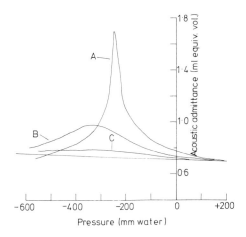

Figure 6.6   Admittance tympanograms at 275 Hz showing low middle ear pressure (A), and middle ear effusions (B and C). The straight line at the base of the diagram represents the admittance of the air in the meatus. For simplicity, the same meatal volume (0·7 ml) has been chosen in each case.

cavity and by increasing the effective mass of the drum. When an effusion is present the admittance peak in the tympanogram is greatly reduced and displaced to the left, as shown in figure 6.6. In many cases a recognisable maximum is absent altogether.

Although tympanometry can distinguish reliably between normal ears and ears with effusions (Brooks 1968, 1969), it often fails to provide a correct diagnosis when other abnormalities are also present. Middle ear scarring, for example, reduces the admittance at the eardrum and when this occurs together with low middle ear pressure the tympanometric pattern is often very similar to that associated with an effusion. The reliability of a tympanometric diagnosis of a middle ear effusion can be estimated from the values of compliance, gradient and middle ear pressure (Grimaldi 1976, Haughton 1977). Compliance and gradient are so well correlated that either quantity may be used as the diagnostic parameter, but middle ear pressure is independent of these quantities and can therefore provide a separate indication of the state of the middle ear. Low intratympanic pressure does not in itself indicate the presence of an effusion, but the finding that middle ear pressure is normal is a reliable indication that an effusion is absent.

Fewer than 2·5% of ears with effusions have middle ear pressue above −50 mm water. When the pressure is less than this the tympanometric diagnosis depends on the value of the compliance or gradient.

It is useful to have a particular value of compliance as a criterion on which to make the decision that an effusion is or is not present. The value of this criterion must depend on the population to which the test is applied and on the failure rate which is to be accepted. When applied to children seen in a typical ENT outpatient clinic, the appropriate criterion is a compliance of 0·21 or a gradient of 0·04 equivalent volume units (Haughton 1977). This gives an average failure rate of about 20% and approximately equal numbers of ears with and without fluid are not correctly identified. For a given ear, the probability that the tympanometric diagnosis is correct depends on the particular value of compliance obtained. Extreme values of compliance are of course a more reliable indication of the presence or absence of effusion than values close to the criterion, and tympanometry is ambiguous when the compliance is in the range 0·18–0·25 ml (gradient 0·03–0·06). Fortunately this range encompasses only about 10% of the cases likely to be seen in the outpatient population referred to. For school screening slightly different criteria might be appropriate. A compliance of 0·35 ml, if used as a criterion, would result in the over-referral of 2·5% of normal ears, while 10% of ears with effusions would be missed. The number of normal ears over-referred would be greatly reduced if a negative middle ear pressure below −50 mm water were an additional requirement for the diagnosis of an effusion.

6.4.3   Otosclerosis

Fixation of the stapes increases the acoustic impedance at the eardrum. The possibility of using impedance measurements as an aid to the diagnosis of otosclerosis was one of the considerations which led to the development of impedance meters for clinical use. Measurements of the compliance at the eardrum in ears with otosclerosis have been reported by Zwislocki and Feldman (1970) and by Jerger *et al* (1974a, b). Although there is a considerable overlap in the distributions of compliance for normal ears and for ears with otosclerosis (table 6.4), impedance tests can often be helpful in the diagnosis of this disease. It is important to note that in this context the impedance measurements are valid only if the appearance of the tympanic membrane is normal. Eardrum abnormalities are usually the result of a middle ear disease unrelated to otosclerosis and as a consequence the eardrum impedance

Table 6.4 Compliance at the eardrum in normal ears and ears with otosclerosis.

	Number of ears, N	Compliance in ml equivalent volume		
		Percentile		
		10	50	90
Normal ears Jerger *et al* (1972)				
220 Hz	1650	0·39	0·67	1·30
Otosclerosis Jerger *et al* (1974)				
220 Hz	95	0·10	0·35	1·01
Otosclerosis Zwislocki and Feldman (1970)				
250 Hz	24	0·15	0·25	0·36

Notes

(*a*) Data for normal ears are also given in table 6.2.

(*b*) In Jerger's measurements compliance was determined at middle ear pressure. See notes (*a*) and (*b*) in table 6.2.

(*c*) Zwislocki and Feldman's measurements were made at ambient atmospheric pressure using a Zwislocki bridge. The meatal volume was determined by filling the ear with alcohol. Resistive and reactive components of impedance were determined separately; the resistive component makes a negligible contribution to the modulus of the impedance at 250 Hz.

may be abnormal regardless of the mobility of the stapes. If eardrum abnormalities are present the clinical diagnosis of otosclerosis is difficult and unfortunately tympanometry is not helpful in these circumstances.

6.4.4 Perforations of the tympanic membrane

At 220 Hz even a small perforation is sufficient to bypass the eardrum impedance. Provided that the middle ear volume has not been abolished by the collapse of the eardrum into the tympanic cavity, the impedance presented to the probe is then due almost entirely to the compliance of the air in the meatus and in the middle ear space 'seen' through the perforation. The tympanogram is therefore a straight line representing a constant high compliance with an equivalent volume usually greater than 2 ml. The presence of a perforation can often be confirmed by operating the pneumatic system of the impedance meter manually and observing the pressure gauge. A positive pressure in the external ear is often relieved suddenly by the spontaneous opening of the Eustachian tube, and this is accompanied by a transient increase in compliance. With a negative pressure, however, the Eustachian tube remains closed unless the patient swallows.

6.4.5　Tympanic membrane scarring and ossicular discontinuity

Scarring of the tympanic membrane, perhaps at the site of a healed perforation, may produce a flaccid area in the eardrum, or the eardrum as a whole may be hypermobile as a result of a break in the ossicular chain. Both these conditions lead to an abnormally high compliance so that the tympanogram has an unusually large peak. At the higher frequencies (660 Hz) the tympanogram is often bimodal ('notched') but the appearance of a double peak does not necessarily indicate a middle ear disorder (Vanhuyse *et al* 1975).

6.5　Middle Ear Reflexes

The action of the middle ear muscles has been the subject of considerable interest both for its own sake and in relation to the diagnosis of hearing disorders. The extensive literature on this topic has been reviewed by Jepsen (1963) and Djupesland (1975). There are principally three methods for examining the aural reflex: electromyography of the tensor tympani and stapedius muscles (Salomon and Starr 1963, Djupesland 1965); observation of displacement of the eardrum (Casselbrant *et al* 1977); and the measurement of acoustic impedance. Of the three, the measurement of impedance has the advantage that it is a simple non-invasive procedure which can easily be incorporated into a diagnostic test.

Contraction of the middle ear muscles increases the stiffness of the ossicular system (§2.4) and the associated decrease in compliance can be detected using a low-frequency (220 Hz) impedance meter. The reduction in compliance varies with the contraction of the muscles, and this in turn depends on the nature of the stimulus. The maximum change in compliance varies considerably from one subject to another, but is typically about 0·1 ml equivalent volume. This is much larger than the real volume change in the meatus produced by movement of the eardrum in response to the action of either the stapedius muscle or the tensor tympani. Casselbrant *et al* (1977) found that contraction of the stapedius muscle produced either inward or outward movement of the drum depending on the subject. The associated change in volume was usually less than $0·5 \mu l$. Occasionally much greater changes in volume were observed, which corresponded to inward movements of the eardrum and were attributed to the action of the tensor tympani. These volume changes were often too large to be measured. Although the tensor tympani pulls the drum inwards increasing the meatal volume, the reflex is nearly always accompanied by a decrease in compliance even in hypermobile ears such as those with ossicular disruption (Klockhoff 1961). Thus it may be concluded that when the middle ear muscles contract, the change in impedance due to the change in the volume between the probe and the eardrum is swamped by the change in impedance due to a modification of the mechanical behaviour of the middle ear. The same is also assumed to be true when eardrum movements are produced by altering the pressure in the external ear during tympanometry.

It is generally believed that the reflex involving the tensor tympani is part of a defensive reaction which occurs when the subject is startled or when he anticipates something unpleasant, such as a loud noise (Djupesland 1975). Unlike the stapedius response, which is sustained for the duration of the stimulus, the tensor tympani response is transient, at least for acoustic stimulation (Djupesland).

Although various stimuli may be used to obtain contractions of the middle ear muscles, the acoustic stimulus is the most widely used and the most important. Generally this stimulus is presented in a way which does not startle the subject or elicit a defensive reaction. The response, which is called the acoustic reflex, provides information relating either to the conductive function in the ear containing the probe or to the auditory function in the ear receiving the stimulus.

6.5.1 Auditory tests involving the acoustic reflex

The stapedius muscles in both ears contract when either ear is exposed to a loud sound. The threshold sound pressure level in the meatus required to elicit the reflex is usually in excess of $80 \, dB$ (re $20 \, \mu Pa$) and on some occasions stimuli up to $120 \, dB$ may be required. The sound pressure in the meatus due to the probe tone of the impedance meter is usually less than 95 dB. It is therefore difficult to present the stimulus and to measure the impedance in the same ear simultaneously, and for this reason the probe is usually placed in the contralateral ear. However, it is necessary that the ear containing the probe is free from any conductive defect which would prevent the reflex being detected and, since this requirement cannot always be met, some instruments are designed to operate with a homolateral (ipsilateral) stimulus. In these instruments the component of the microphone output due to the stimulus has to be removed by a high-quality filter or, alternatively, a multiplex system may be used in which the probe tone and the stimulus are presented alternately in rapid succession.

Hearing tests involving the acoustic reflex nearly always necessitate a determination of the threshold stimulus for the reflex. This is known as the acoustic or stapedius reflex threshold (ART, SRT). It is defined as the least stimulus which produces a detectable change in impedance and it therefore depends to some extent on the sensitivity of impedance measuring instruments. This may account for the differences in threshold intensities reported by various workers (Jepsen 1963, Jerger *et al* 1972, French St George and Stephens 1977). According to Jerger, the normal reflex threshold for frequencies between 500 and 4000 Hz has a Gaussian distribution with a mean and standard deviation of 84.3 and 8.04 dB (ISO), respectively.

Table 6.5 gives acoustic reflex thresholds for ears with cochlear disorders. It can be seen that the reflex threshold is almost independent of the threshold of hearing provided that this is less than about 65 dB. The phenomenon is similar to loudness recruitment in so far as the stimulus required to evoke a given sensation of loudness is also independent of the threshold of hearing if the loudness is sufficiently great. When loudness recruitment is absent the reflex threshold is elevated (Metz 1952).

If the stimulus for the reflex is a band of noise rather than a pure tone, critical-band phenomena may be observed (Flottorp *et al* 1971). Thus if the bandwidth is varied, the reflex threshold (expressed in terms of the overall sound pressure level in the band) remains constant provided that the bandwidth is less than the critical value. Beyond the critical band the reflex threshold decreases linearly with the logarithm of the bandwidth. Critical bandwidths obtained from the reflex threshold are much larger than those obtained using the psychoacoustic techniques mentioned in §3.7. This discrepancy may exist because the middle ear reflex is an involuntary response controlled by a neural

Table 6.5   The acoustic reflex threshold in ears with cochlear lesions.

| Hearing threshold (dB) (ISO) | Reflex threshold level (dB) (ISO) | | | | | |
| | Jerger | | | | | Priede and Coles 1000 Hz |
	500 Hz	1000 Hz	2000 Hz	Mean, 500, 1000 and 2000 Hz	4000 Hz	
0	(85)	(80)	(83)	(82·7)	(83)	(87)
15	85	83	84	84·0	84	88
25	85	83	87	85·0	94	89
35	86	86	87	86·3	89	92
45	85	87	88	86·7	94	95
55	85	90	90	88·3	98	98
65	95	93	95	94·3	103	102
75	96	104	107	102·3	—	108
85	109	—	—	—	—	116

Notes

(*a*) The data in this table are taken from figure 4 in Jerger *et al* (1972) and from figure 8 in Priede and Coles (1974).

(*b*) Jerger's data relate to 515 subjects (1030 ears) with sensorineural hearing losses, presumed to be of cochlear origin. Priede's data relate to 96 ears with cochlear lesions.

(*c*) The patients were grouped according to their hearing loss; the hearing thresholds shown in the table are the median values for each of the groups. The reflex thresholds are also median values.

(*d*) The figures in brackets are extrapolated values.

mechanism at a subcortical level, whereas the psychoacoustic response depends on a conscious evaluation of the stimulus involving high levels of cerebral activity.

Critical bands determined by loudness summation (figure 3.9) are considerably wider in ears with a defective cochlea than in normal ears (Scharf and Hellman 1966), but this enlargement of the band-width does not seem to occur when critical ratios are measured (Scharf 1970). The experiments with loudness summation suggest that the loudness of a given band of noise increases with the number of critical bands falling within its bandwidth. By analogy, the acoustic reflex threshold for wide-band noise should also depend on the number of critical bands within the stimulus. Since cochlear impairment increases the width of the critical bands it reduces the number of bands that are influenced by a given wide-band stimulus. Thus the reflex threshold for this type of stimulus is higher in subjects with cochlear lesions than in subjects with normal ears. This contrasts with the reflex threshold for pure tones which, as stated earlier, is relatively insensitive to cochlear damage. It is therefore possible to estimate the extent of a sensory (cochlear) hearing loss by comparing the acoustic reflex thresholds for pure tones with the thresholds for wide-band noise.

The use of reflex measurements for estimating auditory thresholds may have an application in testing young children and others for whom conventional audiometry is inappropriate. Some attempts have already been made to evaluate this technique, with promising results (Niemeyer and Sesterhenn 1974, Jerger *et al* 1974a, b, Miller *et al* 1976, Keith 1977, Sesterhenn and Breuninger 1977). In the methods reported by these authors the reflex threshold for wide-band noise is subtracted from the average reflex threshold for a number of pure tones at various frequencies (for example, 500, 1000 and 2000 Hz). An empirical relationship between this difference and the hearing loss for pure tones is found by testing normal and hearing-impaired subjects whose hearing thresholds are known. The work so far reported suffers from a certain arbitrariness, particularly in the methods of formulating the average reflex threshold for pure tones. A difficulty is that wide-band noise is a peculiarly ill-defined stimulus when delivered through earphones. The ideal noise would be, perhaps, one in which the spectrum level over a specified bandwidth was constant relative to the normal threshold of hearing or relative to the normal threshold for the reflex. This form of stimulus might be difficult to obtain, partly because the normal threshold of hearing is defined only at a few discrete frequencies, though interpolation between these frequencies might be satisfactory.

An alternative stimulus, which is easier to reproduce than random noise, is a 'synthetic' noise generated by the combination of a series of pure tones with frequencies centred on the normal critical bands. Niemeyer and Sesterhenn (1974) found that this type of noise gave similar results to those obtained with random noise.

6.5.2 The use of the stapedius reflex in the diagnosis of conductive disorders

Klockhoff (1961) was impressed by the finding that only a trivial conductive abnormality was needed to prevent his detecting the change in impedance associated with a contraction of the stapedius muscle. He considered the presence of a demonstrable reflex to be almost conclusive proof of normal conductive function. Nowadays, many audiologists would think this to be an over-simplification; the ability to detect the reflex depends on the impedance meter, and sensitive instruments can often detect a reflex when abnormalities are known to exist. Although there is no published information on the magnitude of the impedance change that could be regarded as normal, clinical experience shows that the change in impedance is abnormally small when a conductive lesion is present. Thus the appearance of a 'normal' stapedial reflex can usually be considered good evidence for normal middle ear function.

The absence of an acoustic stapedial reflex may be due to a conductive lesion in the ear containing the probe or a hearing loss in the ear receiving the stimulus. If the deafness is sensorineural the absence of a reflex may be evidence for a neural lesion and it may therefore be important to demonstrate that, but for the deafness, a reflex would have been obtained. Neural lesions are usually unilateral so that in these cases a normal reflex in the ear containing the probe may be demonstrated with the aid of a homolateral acoustic stimulus if the facility for presenting it is available. The alternative is to attempt to elicit the reflex with non-acoustic stimuli as described by Klockhoff (1961), Jepsen (1963) and Djupesland (1964, 1975). The simplest of these stimuli is tactile, involving the application of a cotton wool 'twist' to the skin of the face on the same side as the ear in which the impedance change is detected. A stapedius reflex can also be evoked by directing a jet of air to either of the external ears (Djupesland 1961, 1962), or by electric stimulation of the skin in the homolateral auditory canal (Klockhoff 1961).

As already stated, contraction of the middle ear muscles normally increases the acoustic impedance of the ear at low frequencies. However, with some middle ear disorders a decrease in impedance can occur, or the impedance change may be transient although the contraction of the muscle is sustained. The most noteworthy occurrence of this phenomenon is in ears with incipient otosclerosis. In this condition a transient decrease in impedance is observed when the stimulus is presented and again when it is removed (Flottorp and Djupesland 1970). An explanation of this so-called 'diphasic' response has been suggested by Bel *et al* (1976).

7 Hearing Aids

7.1 Introduction

The reception of speech depends principally on auditory performance in the mid-frequencies (500–2000 Hz). A small hearing loss in these frequencies can usually be tolerated but the understanding of normal speech is impaired once the loss exceeds about 30 dB. The main reason for this is simply that when a hearing loss is present, ordinary speech may be received at a sensation level below that necessary for correct understanding. This can be seen from the speech audiograms (figure 5.5) which show that the principal effect of deafness is a displacement of the intelligibility function to higher relative speech levels. A second consequence of deafness, if it is sensorineural, is that complete intelligibility is not achieved at any speech level.

An ideal hearing aid should therefore perform two functions: it should amplify speech signals so that they are received at an adequate level for understanding and, for users with sensorineural lesions, it should modify these signals in a way that compensates for the impairment of the analytical capacity of the ear. The first of these requirements is not difficult to meet, but the second has remained an almost intractable problem.

A hearing aid may be regarded as part of a speech transmission system but it differs from comparable systems used in telecommunications in two respects. The first of these is that the communication is between a normal speaker and a hearing-impaired listener; the second is that a hearing aid is a wearable device which, for cosmetic reasons, is usually made as small as practically possible. When considering what electroacoustic characteristics might be appropriate for a hearing aid it is relevant to examine first the acoustic properties of speech and the factors affecting its intelligibility for listeners with normal hearing. After this it is necessary to consider what additional factors influence intelligibility when the recipients do not have normal hearing. Then, having decided what performance is required, the final problem is the technical one of constructing a device which has the requisite electro-

148

acoustic characteristics and which can be mass-produced cheaply from miniature components.

7.2   Physical Characteristics of Speech

The physical characteristics of speech and the mechanism of voice production have been the subjects of numerous investigations. A resumé of much of the early work in this field is provided by Littler (1965) and very much more detailed accounts are given by Fletcher (1953) and Flanagan (1972).

Speech is excited by the flow of air from the lungs. In voiced speech this flow is interrupted by the vocal folds which produce the so-called laryngeal tone, but in unvoiced speech this mechanism is not involved. The vocal folds open and close at rates ranging from 75 to 500 per second depending on intonation and voice quality. The average number of interruptions per second is about 120 for males and 240 for females. The frequency spectrum of speech depends on the laryngeal tone and on resonances within the vocal cavities (throat, mouth and nasal cavities) which are continually modified in the act of speaking. Resonances in the vocal tract reinforce particular frequency bands of laryngeal tone, and for vowels and diphthongs several characteristic frequencies known as formants can be identified. The bandwidth in which resonance occurs decreases with increasing duration of the sound, so that the longest vowels may be associated with bandwidths of only a few hertz. Littler (1965) describes syllables as vowels to which transients and transitional sounds—the consonants—have been added. The consonants involve the production of turbulence at some point in the vocal tract, either by forcing the exhaled breath through a constriction or by completely blocking the air flow and then suddenly releasing it. In acoustic terms, however, vowels and consonants are not clearly differentiated because some consonants have a sustained or periodic character.

The frequency characteristics of particular phonemes or words may be examined with the help of a sound spectrograph (Koenig *et al* 1946, Potter *et al* 1947). The spectrograph samples speech, a few words at a time, and leaves a visible record showing how the sound energy in a series of specified frequency bands varies from one instant to the next. Another form of analysis gives the average power spectrum in a passage of speech which is long enough to include a large number of speech sounds. The long-term speech spectrum (figure 7.1) is often expressed as the spectrum level per hertz although, as pointed out by Byrne (1977), this is a rather misleading representation. Since frequency is conventionally shown on a logarithmic scale it is more logical to express the spectrum level in terms of the energy per unit relative bandwidth, that is, in dB per octave or $\frac{1}{3}$-octave. Compared to the level per hertz, the latter rises with frequency at 3 dB per octave. Byrne has suggested that the psychophysical critical bands might be appropriate bands in which to sample the speech signal, but as the spectrum levels for critical bands are close to those for $\frac{1}{3}$-octave bands this procedure would be of theoretical rather than practical significance.

The phonemes which make up a passage of speech differ in their relative intensities over a range of almost 30 dB (table 7.1 and figure 7.2) so that the acoustic power in continuous speech varies considerably from one instant to the next. The instantaneous power is of course zero during the many pauses which occur between words and phrases and even within individual words themselves. The average power

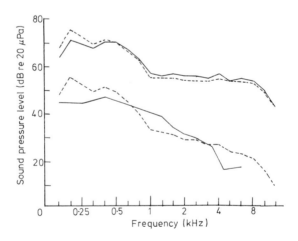

Figure 7.1 The long-term speech spectrum. The ordinate is the average rms free-field pressure level. Upper curves: pressure level expressed in $\frac{1}{3}$-octave bands (full line) or critical bands (broken line). Lower curves: spectrum level per hertz. The upper curves and the lower broken curve (taken from Byrne 1977) represent an average for 15 male and 15 female Australian speakers; the lower full curve (from Rudmose *et al* 1948) is an average for seven American males.

Table 7.1 The powers of speech sounds (in dB) relative to the power of the weakest sound. See also figure 7.2. From *Speech and Hearing in Communication* by Harvey Fletcher © 1953 by Litton Educational Publishing, Inc. Reprinted by permission of Van Nostrand Reinhold Company.

talk	ó	28·3	err	r	23·2	tap	t	11·8
top	a	27·8	let	l	20·0	get	g	11·8
ton	o	27·1	shot	sh	19·0	kit	k	11·1
tap	á	26·9	ring	ng	18·6	vat	v	10·8
tone	ō	26·7	me	m	17·2	that	<u>th</u>	10·4
took	u	26·6	chat	ch	16·2	bat	b	8·5
tape	ā	25·7	no	n	15·6	dot	d	8·5
ten	e	25·4	jot	j	13·6	pat	p	7·8
tool	ū	24·9	azure	zh	13·0	for	f	7·0
tip	i	24·2	zip	z	12·0	thin	th	0·0
team	ē	23·4	sit	s	12·0			

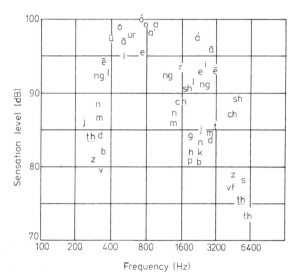

Figure 7.2 Approximate frequency and intensity relations of the speech sounds listed in table 7.1. The ordinate shows the relative sensation levels of the principal frequency components of each sound. The level of the strongest sound is arbitrarily set to 100 dB. Where a sound has several frequency components the position of each is indicated. From Speech and Hearing in Communication by Harvey Fletcher. Copyright © 1953 by Litton Educational Publishing, Inc. Reprinted by permission of Van Nostrand Reinhold Company.

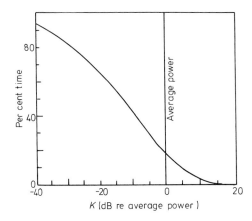

Figure 7.3 Distribution of the instantaneous speech power. The ordinate shows the proportion of time that the instantaneous power is greater than the average power by K dB or more. From Littler (1965), based on data from Sivian (1929).

radiated by the speaker during normal conversational speech is typically about $20\,\mu$W, but the range from a whisper to the loudest possible level is from 10^{-3} to $10^{3}\,\mu$W. If conversational speech is sampled randomly it is found that the instantaneous power is distributed with a bias in favour of the weaker sounds, the average power being exceeded only 20% of the time (figure 7.3).

7.3 Factors Affecting the Intelligibility of Artificially Transmitted Speech

The variation of intelligibility with intensity has already been mentioned in connection with speech audiometry. Other factors affecting the intelligibility of artificially transmitted speech are distortion and the bandwidth of the transmitting system. The effect of reducing the bandwidth has been investigated by French and Steinberg (1947) and Hirsh *et al* (1954). In their experiments, intelligibility was measured when the speech signal was degraded by the insertion of either high- or low-pass filters. Some of the results obtained by Hirsh and his colleagues are shown in figure 7.4. These results are for ordinary phonetically balanced monosyllabic words†; generally lower scores were

† Central Institute for the Deaf, Auditory Test W-22 (Hirsh *et al* 1952).

obtained for meaningless monosyllables ('logatoms') in agreement with the findings of French and Steinberg who also used meaningless words. Much of the information in ordinary speech is redundant and 50% intelligibility for monosyllables corresponds to approximately 90% intelligibility for sentences (Licklider 1946). Thus figure 7.4 may be interpreted as showing that 'satisfactory' communication remains possible, at least with normal listeners, even when the speech signal is degraded by removal of all components with frequencies either below 3 kHz or above 800 Hz. In general, the effect of filtering on intelligibility can be estimated by calculating for the system concerned a quantity called the articulation index (AI) which bears a known relationship to the intelligibility of various forms of spoken material (French and Steinberg 1947, Kryter 1962, ANSI S3.5: 1969).

The bandwidth of any transmitting system is of course finite. Sinusoidal signals outside the pass band are merely attenuated, but complex signals are distorted by an amount depending on the significance of the Fourier components lost in the transmission process. This

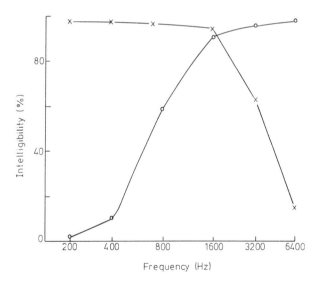

Figure 7.4 The effect of high-pass filtering (—×—) and low-pass filtering (—○—) on the intelligibility of speech. The proportion of words correctly identified is shown on the ordinate. The abscissa shows the cut-off frequency of the filter. From Hodgson (1977), based on data from Hirsh *et al* (1954).

is however only one of several forms of distortion that needs to be considered.

Distortion can be defined for a system having an input and an output as a failure of that system to reproduce in the output the same waveform as that which is present in the input. A difficulty in applying this definition to a system for transmitting speech is that input and output are not easily specified. When the system is a hearing aid, the widest definition of distortion is a failure, in response to a given sound field, to reproduce the same pressure waveform at the eardrum as the waveform which would exist in the unaided ear exposed to the same sound field.

For sinusoids, distortion occurs when the amplitude of the output is not a linear function of the amplitude of the input. This form of distortion leads to the production of harmonics and intermodulation products of the input signal. If the input is a sinusoid at frequency f_1, harmonic distortion is the appearance in the output of components with frequencies $2f_1$, $3f_1$, and so on. If the input is a complex of two sinusoids, f_1 and f_2, the intermodulation products are sinusoids at frequencies $nf_1 \pm mf_2$, where n and m are integers. Another form of distortion, known as transient distortion, is the existence of transients in the amplitude of the output following abrupt changes in the amplitude of an otherwise steady input. A particular example is 'ringing,' which is the persistence of an alternating output after the abrupt termination of the input. Another example is the occurrence of transients following an overload; in some instances the output may be temporarily reduced to zero during recovery from this condition. Transient distortion can occur in all systems whether the steady state performance is linear or not. Since speech is essentially a series of transients it is particularly important that this form of distortion should be avoided in systems used for speech transmission.

The effect of distortion on the intelligibility of speech is difficult to assess, largely because of the complexities of speech signals and the numerous different forms of distortion. Niemoeller *et al* (1970), commenting on this problem in connection with the design of hearing aids, wrote:

> The relation between various measures of distortion in a hearing aid and the ability of an impaired listener to receive information is not fully understood. The problem is too complicated; there are too many kinds of signals of interest, too many kinds of non-linearity and too many kinds of hearing impairment.

An important aspect of verbal communication is that the intelligibility of speech is resistant to changes in the waveform of speech signals. Such changes are the deliberate or accidental consequence of limitations imposed by the transmitting system. In hearing aids the most extensive modification of waveform results from the action of output limiting devices of which the most commonly used is the peak clipper. Taken to extremes, peak clipping converts the speech signal into a rectangular wave whose rising and falling edges are generated whenever the speech wave crosses the time axis. Peak clipping will be considered later, but it is interesting to note here that the intelligibility of monosyllables subjected to infinite clipping (converted to rectangular waves) is at least 70%, and may be as high as 90% once the listener has become accustomed to this form of speech. Moreover, the rectangular wave may be differentiated so that the listener is presented with a series of pulses, yet even this gross modification has little effect on intelligibility. This once more demonstrates the redundancy in speech signals and the ability of the brain to extract meaningful information under difficult conditions.

An interesting account of experiments on the intelligibility of modified speech is given in papers by Licklider (1946) and Licklider and Pollack (1948). From these experiments Licklider concluded that intelligibility depends principally on the temporal coding of the speech signal and that the fine structure of the waveform is of secondary importance.

This is consistent with the observations mentioned earlier on the effects of the low-pass filtering, for if the whole of the speech signal is subjected to infinite clipping much of the information originally present in the weak high-frequency components will be lost. However, some modifications of the speech wave are not well tolerated. Centre clipping, for example, which leaves only the peaks of the original signal, has a disastrous effect on intelligibility. Infinite centre clipping reduces a sinusoid to a series of alternate pulses. In contrast to the high intelligibility of pulses produced by peak clipping and differentiation, the pulses produced by severe centre clipping are completely unintelligible. It may be relevant that for frequencies above 200 Hz the periodicity pitch of the centre-clipped signal probably depends on the fundamental period of the original waveform, namely, the period between successive pulses of the same polarity, but below 200 Hz it is determined by the interval between alternate pulses (Scharf 1970, Flanagan and Guttmann 1960). Thus centre clipping leads to severe distortion of pitch with the lower tones being perceived an octave too high. Whether this accounts for the loss of intelligibility with centre clipping is conjecture, but it is interesting that Licklider observed a similar loss of intelligibility when speech signals were passed through a full-wave rectifier. This form of rectification produces a frequency doubling of the low-frequency components. Half-wave rectification, which does not double the frequency, had relatively little effect on intelligibility.

The harmonic distortion† associated with infinite peak clipping is 43% and 98% when clipping is followed by differentiation. The fact that intelligibility remains high in spite of the severe harmonic distortion exemplifies the statement made earlier that there is no direct relationship between these two quantities.

7.4 Perception of Speech by the Hearing-impaired

Conductive deafness, if it is uniform, merely attenuates the speech signal without altering its waveform. The corresponding reduction in intelligibility can be determined from the normal intelligibility function shown on a speech audiogram (figure 5.5). In conductive deafness this function is displaced, without change in shape, to higher speech levels by an amount equal to the hearing loss for pure tones. Even if a conductive loss is not uniform, so that the speech signal is attenuated more at some frequencies than at others, the rate of change of threshold with frequency is generally small and the effects of selective attenuation are unimportant. Thus intelligibility can easily be restored by amplification.

In contrast to the relatively simple consequences of conductive deafness, cochlear disorders are often disabling and difficult to remedy. Two features of sensorineural impairment particularly detrimental to the perception of speech are high-tone loss and recruitment. A high-tone loss can often be severe and highly selective so that its effect is analogous to low-pass filtering. Amplification of the high tones may improve intelligibility but in these circumstances recruitment is a handicap because the dynamic range of the impaired ear may not be sufficient to accommodate the range of intensities in speech signals.

---

† Per cent harmonic distortion is given by $100\sqrt{(S - p_1^2)/S}$, where p_1 is the sound pressure of the fundamental, p_n is the corresponding sound pressure of the nth harmonic, and $S = \Sigma_1^\infty p_n^2$.

The stronger components of speech are perceived at a level which is uncomfortably loud, while the weaker components are not heard at all. Littler (1965) has likened recruitment to centre clipping but the analogy, though interesting, needs to be treated with caution. Loudness recruitment is a distortion of perception and not a distortion of the physical processes which precede it.

The increase in the auditory threshold and the presence of loudness recruitment are in themselves probably not sufficient to account for the frequent failure of amplification to make speech intelligible to people with sensorineural disorders, and other abnormalities need to be considered. These are described by Stephens (1976) as abnormalities of frequency and temporal coding. Frequency coding abnormalities relate to a reduction in the selectivity of the 'tuning' mechanism of the cochlea and an increase in the width of the critical bands (Kiang *et al* 1970, Evans 1976, Scharf and Hellman 1966). Temporal coding abnormalities are abnormalities in the way the acoustic energy, or the neural information derived from it, is integrated in the auditory system. Stephens suggests that abnormalities of integration may distort the perception of transients so that these receive an undue emphasis relative to steady signals. Other temporal abnormalities which have been reported are an increase in the degree and duration of backward masking and possibly also forward masking (Elliott 1975), and a diminished ability on the part of the listener to detect a brief silent interval in otherwise continuous noise.

7.5 The Optimum Performance of Hearing Aids

The three most important characteristics of a hearing aid are its maximum acoustic output, its gain and its frequency response. The output must be such that as far as possible speech sounds are made audible, but it needs to be limited to avoid causing pain or injury if the aid is inadvertently used at high amplification in noisy surroundings. The maximum output therefore has to be selected with regard to the comfort and tolerance of the individual user. The gain is usually adjusted by the user and will normally be set for comfortable listening. The optimum frequency response is a more difficult characteristic to specify and one that has been the subject of much research.

Frequency response is the function which describes the way in which gain depends on frequency. It is normally measured at some point in the linear region of the input–output characteristic. At first sight it might seem that the appropriate response would be one that compensated for the audiometric hearing loss so that normal free-field thresholds would be obtained with the aid in operation. This view, however, is mistaken because it takes no account of loudness recruitment or the relative importance of different parts of the speech spectrum. In the case of a high-tone loss, threshold compensation may give excessive amplification at high frequencies. Experiments to determine the optimum frequency response were reported by the Medical Research Council (1947) and independently by a group at Harvard University (*Harvard Report*: Davis *et al* 1946, 1947). Although this work was done over 30 years ago, it has had an important influence on the design of hearing aids. The characteristics recommended by the MRC are shown in figure 7.5. The Harvard Report recommended either a uniform response or one that sloped upward at 6 dB per octave to accentuate the high tones.

The frequency response recommended by the MRC is an idealised version of a response derived theoretically from data obtained in the development of communication systems. Although such systems are

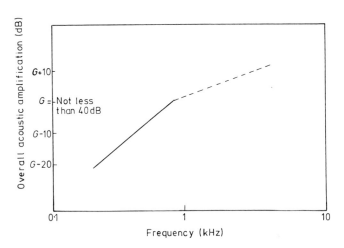

Figure 7.5 Frequency response characteristics for hearing aids as recommended by the Medical Research Council. The ordinate is the free-field to coupler gain, *G*, corrected for diffraction by the head. The broken line is an alternative response to be provided by adjustment of a pre-set tone control. From MRC Special Report Series No. 261 (1947): *Hearing Aids and Audiometers*.

intended for persons with normal hearing it was felt that the analysis would give a reasonable indication of the requirements of the hearing-impaired, and this was confirmed by speech articulation tests with deaf subjects. The theoretical treatment is too lengthy to reproduce here but it is given in an appendix to the MRC report. It is based on the assumption that the total acoustic power of the aid is a fixed quantity, independent of the frequency response of the aid, and determined in practice by the onset of discomfort. The auditory spectrum is divided into a series of frequency bands, each of which contributes to the intelligibility of the transmitted speech by an amount depending on the bandwidth and the location of the band in the frequency scale. It is assumed that intelligibility is severely limited by the hearing loss and that the useful power in each band is likely to be 35 dB or less above the threshold for the band. A calculation is made to determine how the total power must be distributed among the bands in order to maximise intelligibility. The optimum frequency response is then the one which gives this distribution. The calculation can be made for any form of hearing loss: the MRC report gave results for a uniform loss and for a loss of high tones above 1 kHz. An interesting feature of this analysis is that the theoretical response curve for the high-tone loss lies below that for the uniform loss, showing that an aid suitable for high-tone deafness should provide *less* amplification at high frequencies than an aid designed to compensate for a uniform loss. The reason for this is that, for a given total power, it is more profitable to distribute most of the energy to those frequencies which carry the greatest information rather than attempt to transmit the high frequencies which are rela-tively unimportant.

The frequency response shown in figure 7.5 was intended for hearing aids employing the now obsolete external receivers. The amplification shown on the ordinate is the free-field to coupler gain measured with the microphone facing a source of plane progressive waves and the earphone attached to an artificial ear. The MRC report included, however, a correction for diffraction by the head, and the inference is that the authors of the report intended their specification to represent the functional gain at each frequency. Functional gain may be defined as the ratio of the sound pressure at the eardrum when the aid is in use to the corresponding sound pressure when the aid is removed. The amplification shown in figure 7.5 differs from the functional gain as a result of deficiencies in the artificial ear and the modifying influence of the earphone on the acoustic characteristics of the outer ear. As a consequence, the proposed correction for diffraction by the head may have been inadequate at some frequencies. Nevertheless the MRC specification provides a reasonable description of the optimum functional gain for a mass-produced instrument. The specification is restricted to frequencies in the range 200–4000 Hz since both the MRC and the Harvard experiments showed that no benefit would be obtained by providing amplification at frequencies outside this range.

The Harvard and MRC recommendations have recently been criticised by Pascoe (1975) on the grounds that they do not relate to modern hearing aids employing ear level microphones and insert receivers. Experiments using such microphones and receivers gave results which seemingly contradicted the earlier conclusions, but a careful examination of Pascoe's results suggests that they are in fact consistent with the principles established by the MRC. Pascoe used an experimental (master) hearing aid which enabled various response characteristics to be selected and assessed in terms of the efficiency of the aid in transmitting speech to deaf subjects. The best results were obtained when the aid had a frequency response designated 'uniform hearing level' in the range 250–6300 Hz. This response was the inverse of the subject's audiogram so that his aided thresholds, when plotted against frequency, formed a curve running parallel to the normal threshold of hearing. Since the average hearing loss of the subjects employed in these experiments increased with frequency, the average response needed to give 'uniform hearing level' had also to increase with frequency, and of the various responses tested this configuration came, perhaps fortuitously, closest to the MRC specification. The significance of extending the operating range beyond 4 kHz is not clear since this modification was not tested independently. Pascoe found that intelligibility with an aid giving a 'uniform hearing level' was greatly superior to that obtainable with an aid having the frequency response of a typical commercial instrument. When tested on a 2 cc coupler the latter had a response similar to that recommended by the MRC, but in functional terms the coupler measurements under-estimated the gain in the range 1–2 kHz and seriously over-estimated it in the range 2–4 kHz. Thus the relatively poor performance of the commercial aid could be attributed mainly to inadequate amplification at frequencies above 2 kHz.

A final consideration in regard to frequency response is that the response curve should as far as possible be smooth. According to the MRC and Harvard reports, rapid fluctuations in frequency response are likely to have an adverse effect on intelligibility, and this has been confirmed by Jerger and Thelin (1968).

7.6   Output-limiting Techniques

The maximum acoustic output of a hearing aid is usually controlled either by peak clipping or by the action of an automatic gain control. As discussed earlier, peak clipping is not seriously detrimental to intelligibility, at least when the speech is received by someone with normal hearing. Excessive peak clipping makes speech harsh and unnatural and there is of course no virtue in burdening the recipient with any unnecessary reduction in the quality of the transmitted sound. The effects of peak clipping are mitigated if it is restricted to components of the speech signal above about 2 kHz. Harmonics generated by clipping will then be outside the pass band of the aid and therefore inaudible. Conversely, it is undesirable that low-frequency components should be clipped preferentially since this will generate audible harmonics and also distort or suppress the high-frequency components which are superimposed on the low-frequency waveform. It is therefore advantageous to establish the rising portion of the frequency response at a stage which precedes the output limiter so that the high-frequency components are emphasised and clipped preferentially.

As an alternative to peak clipping, the output can be limited by an automatic gain control (AGC). This is often called 'compression', there being no clear distinction between these two terms. In this form of amplification a portion of the output is rectified and integrated to obtain a direct voltage proportional to the amplitude of the output. This voltage is the source of a negative feedback which controls the gain of the amplifier. The action of the gain control is not necessarily uniform over the dynamic range of the amplifier and the use of compression is sometimes restricted to high output levels in order to achieve what is known as 'peak-limiting' (figure 7.6). This differs from peak clipping by preserving the waveform of high-amplitude signals. In the absence of automatic gain control, a linear amplifier responds to an increase (n dB) in the level of the input by generating the same decibel increase in the output. With automatic gain control the change in output level (m dB) is less than the corresponding change in the input. The ratio n/m is known as the compression ratio of the amplifier. Thus in a peak-limiting circuit the compression ratio is unity up to some

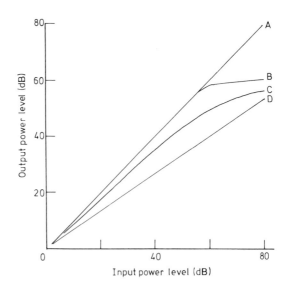

Figure 7.6 Input–output characteristics of a hearing aid with various forms of automatic gain control (AGC). The average input and output power levels for steady signals are shown in dB on arbitrary scales. A, linear amplifier without ACG; B, peak limiting; C, non-uniform compression; D, constant compression.

specified output level, and much greater than unity thereafter. The alternative arrangements represented in figure 7.6 have either a constant compression ratio or one which increases progressively with the strength of the signal.

The chief advantage that automatic gain control has over peak clipping is that the output limitation is achieved without gross distortion of the signal. A second advantage is that if the external volume control is placed after the feedback circuit, the output can be adjusted to a comfortable level which is then maintained regardless of the level of the input so long as this is sufficient to bring the automatic control into operation. A disadvantage of automatic gain control is that it is a source of transient distortion. The distortion is characterised by two time constants known as attack time and recovery (decay or release) time. An abrupt change in the level of an otherwise steady input signal is accompanied by an abrupt change in the amplitude of the output, causing it to overshoot the steady state value. The output is restored to the steady level at a rate determined by the time constant for integration of the output signal. The attack time is the time constant associated with an increase in the input, and the recovery time is the corresponding time constant for a decrease in the input. These time constants are not necessarily equal. To protect the user from the unpleasant effects of a sudden overload, the attack time has to be less than 20 ms (Johansson 1973) but the recovery time has to be much longer than this in order to provide stable amplification and to smooth out fluctuations in gain from one syllable to the next.

The relationship between the parameters of automatic gain control and the intelligibility of amplified speech has been investigated by several workers including Lynn and Carhart (1963), Caraway and Carhart (1967) and Fleming and Rice (1969). The consensus seems to be that compression is often only marginally superior to peak clipping in preserving the intelligibility of strong signals but that it offers some advantages in terms of comfort.

7.7 Measurement and Specification of Hearing Aid Performance

The gain of a hearing aid and the extent to which it enhances the intelligibility of speech can be measured by comparing the aided and unaided performances of a listener in free-field conditions. This form of testing can be used to obtain accurate measurements of the aid but it is not suitable for measuring other physical characteristics such as output power and distortion. Moreover, it is exceedingly time-consuming. A very much simpler procedure is to examine the output of the aid when its earphone·(receiver) is connected by an acoustic coupler or artificial ear to a measuring microphone. However, a simple objective test which accords accurately with the performance on a real ear has yet to be devised and coupler measurements, though quick and highly reproducible, are often seriously lacking in this respect.

The ideal artificial ear should present the earphone under test with the same acoustic impedance as that of the real ear; its microphone should register the sound pressure which would exist at the real eardrum. It is also desirable that this sound pressure is reproduced when the artificial ear is exposed to a free sound field, and to simulate meatal resonance and diffraction under free-field conditions the artificial ear can be placed in a dummy head. Artificial ears which have these desirable properties have been constructed for laboratory use (Brüel *et al* 1977) but are not yet available commercially. The most widely used coupler is a 2 cc cavity described in BS 3171 and IEC 126, and methods for measuring hearing aid performance using this device are described in several of the standards listed in the Appendix (IEC 118, BS 3171 and ANSI S3.8).

7.8 Recoded Speech

Systems known as 'vocoders' (voice coders) have been developed for telecommunications. These systems analyse the speech input, transmit the results of this analysis and, at the receiving end, perform a synthesis which reconstructs the input. In the process much of the redundant information present in the original speech is discarded in order to minimise demands on the capacity of the transmitting system. There are many forms of vocoder and a description of these is to be found in the book by Flanagan (1972). The principle of voice coding can be illustrated by reference to a system known as a spectrum channel vocoder. At the analytical stage of this device, the short-term amplitude spectrum of the speech input and the mode of excitation (voiced or unvoiced) are determined. This information is transmitted to the receiver where it is used to control the production of artificial speech. The artificial speech is excited by one of several sources selected according to the mode of excitation of the original input. The original spectrum is reconstructed in accordance with the information received from the analyser by passing the output of the appropriate source through a set of filters and attenuators which mimic the action of the vocal tract. The output of each of the filters is combined and the resultant signal passed to a loudspeaker. The importance of this process for telecommunications is that the transmission system has only to carry the information needed to control the production of synthetic speech, and since this is considerably less than the 'information' present in the original speech signal, a saving in bandwidth is accomplished.

Voice coders for telecommunications are usually designed to produce a good copy of the original speech; the synthetic product differs from the original only to the extent of the economies to be effected in transmission. However, voice coders can be used as a means of deliberately altering speech, for example, by transposing its spectrum to lower frequencies. This raises the possibility that alterations might be made which would render speech more intelligible to a deaf person (Risberg 1969). Hearing aids employing the techniques used in vocoders have been constructed but the results are not encouraging. Possible reasons for this failure are given in an interesting article by Roworth (1970) in which the particular obstacles to creating successful hearing aids for severely deaf people are discussed. Roworth points out that language is developed at a very early age and that it cannot be acquired later in life. A hearing aid has to have an output which can be recognised as having a linguistic significance, and thus a recoding device which produced a major alteration to the speech code would have to be worn in infancy when, hopefully, its output would be interpreted as speech.

Appendix: Standards Relating to Audiology

ISO 389	(1975)	Acoustics—Standard reference zero for the calibration of pure tone audiometers.
ISO 532	(1975)	Acoustics—Methods for calculating loudness.
ISO 1999	(1975)	Assessment of occupational noise exposure for hearing conservation purposes.
IEC 118	(1959)	Recommended methods for measurements of the electro-acoustical characteristics of hearing aids.
IEC 126	(1961)	Reference coupler for the measurement of hearing aids using earphones coupled to the ear by means of ear inserts.
IEC 177	(1965)	Pure tone audiometers for general diagnostic purposes.
IEC 178	(1965)	Pure tone screening audiometers.
IEC 303	(1970)	IEC provisional reference coupler for the calibration of earphones used in audiometry.
IEC 318	(1970)	An IEC artificial ear, of the wide band type, for the calibration of earphones used in audiometry.
IEC 373	(1971)	An IEC mechanical coupler for the calibration of bone vibrators having a specified contact area and being applied with a specified static force.
BS 661	(1969)	Glossary of acoustical terms.
BS 2042	(1953)	An artificial ear for the calibration of earphones of the external type. Amendment PD 1795 January 1954.
BS 2497	(1954)	The normal threshold of hearing for pure tones by earphone listening.

BS 2497†		Reference zero for calibration of pure tone audiometers.
Part 1	(1968)	Data for earphone coupler combinations maintained at certain standardising laboratories.
Part 2	(1969)	Data for certain earphones used in commercial practice. Amendment AMD 705 March 1971.
Part 3	(1972)	Data for wideband artificial ears complying with BS 4669.
Part 4	(1972)	Normal threshold of hearing for pure tones by bone conduction.
BS 2980	(1958)	Pure tone audiometers. Agrees with IEC 177.
BS 3045	(1958)	The relation between the sone scale of loudness and the phon scale of loudness level.
BS 3171	(1959)	Characteristics and method of test of air conduction hearing aids using electronic amplification.
BS 3171	(1968)	Methods of test of air conduction hearing aids. Agrees with IEC 118.
BS 3383	(1961)	Normal equal-loudness contours for pure tones and normal threshold of hearing under free-field listening conditions.
BS 4009	(1975)	An artificial mastoid for calibration of bone vibrators used in hearing aids and audiometers. Agrees with IEC 373.
BS 4198	(1967)	Method for calculating loudness.
BS 4668	(1971)	An acoustic coupler (IEC reference type) for calibration of earphones used in audiometry. Agrees with IEC 303.
BS 4669	(1971)	An artificial ear of the wide band type for the calibration of earphones used in audiometry.
BS 5108	(1974)	Method of measurement of attenuation of hearing protectors at threshold.
BS 5330	(1976)	Method of test for estimating the risk of hearing handicap due to noise exposure.
ANSI S3.1	(1960)	Criteria for background noise in audiometer rooms.
ANSI S3.4	(1968)	Procedure for the computation of loudness of noise.

† This standard replaces the 1954 version.

ANSI S3.5	(1969)	Methods for the calculation of the articulation index.
ANSI S3.6	(1969)	Specifications for audiometers.
ANSI S3.7	(1973)	American National Standard method of coupler calibration of earphones.
ANSI S3.8	(1967: R1971)	American National Standard method of expressing hearing aid performance.
ANSI S3.13	(1972)	An artificial headbone for the calibration of audiometer bone vibrators.

The above publications are obtainable from the British Standards Institution, 2 Park Street, London W1A 2BS.

ANSI—American National Standards Institute, 1430 Broadway, New York 10018, USA.

ISO—International Organisation for Standardisation.

IEC—International Electrotechnical Commission (affiliated to the ISO), Bureau Central de la Commission Electrotechnique Internationale, 1 rue de Varembé, Geneva, Switzerland.

References

Anderson C M B and Whittle L S 1971 Physiological noise and the missing 6 dB *Acustica* **24** 261–72

Ballantine S 1928 Effect of diffraction around the microphone in sound measurements *Phys. Rev.* **32** 988–92

von Békésy G 1947 A new audiometer *Acta Otolaryngol.* **35** 301–15

—— 1955 Human skin perception of travelling waves similar to those on the cochlea *J. Acoust. Soc. Am.* **27** 830–41

—— 1960 *Experiments in Hearing* ed E G Wever (New York: McGraw-Hill)

—— 1961 Concerning the fundamental component of periodic pulse patterns and modulated vibrations observed on the cochlear model with nerve supply *J. Acoust. Soc. Am.* **33** 888–96

Bel J, Causse J, Michaux P, Cézard R, Canut Y and Tapon J 1976 Mechanical explanation of the on-effect (diphasic impedance change) in otospongiosis *Audiology* **15** 128–40

Bluestone C D, Beery Q C and Paradise J L 1973 Audiometry and tympanometry in relation to middle ear effusions in children *Laryngoscope* **83** 594–604

Boothroyd A 1968 Developments in speech audiometry *Sound* **2** 3–10

Borg E 1976 *Acoustic Impedance and Admittance* ed A S Feldman and L A Wilber (Baltimore: Williams and Wilkins) ch 11

Brindley F J 1974 *Medical Physiology* ed V B Mountcastle (St Louis: C V Mosby) (a) ch 2: Excitation and conduction in nerve fibers; (b) Volume conduction theory pp 247–53

Brödel M 1940 *The 1940 Year Book of the Eye, Ear, Nose and Throat* (Chicago: The Year Book Publishers Inc) p 339

Brooks D N 1968 An objective method of detecting fluid in the middle ear *Int. Audiol.* **7** 280–6

—— 1969 The use of the electro-acoustic impedance bridge in the assessment of middle ear function *Int. Audiol.* **8** 563–9

—— 1971 Electroacoustic impedance bridge studies on normal ears of children *J. Speech Hearing Res.* **14** 247–52

—— 1973 Hearing screening: a comparative study of an impedance method and pure tone screening *Scand. Audiol.* **2** 267–72

—— 1974 The role of the acoustic impedance bridge in paediatric screening *Scand. Audiol.* **3** 99–104

—— 1976 School screening for middle ear effusions. *Ann. Otol. Rhinol. Laryngol.* **85** suppl. 25 223–8

Brüel P V, Frederiksen E *et al* 1977 *Impedance of Real and Artificial Ears* (Naerum, Denmark: Brüel and Kjaer)

Brüel P V and Kjaer 1967 *Instruction and Applications, Microphone Calibration Apparatus Type 4142* Anon. (Naerum, Denmark: Brüel and Kjaer)

Burns W and Robinson D W 1970 *Hearing and Noise in Industry* (London: HMSO)

Byrne D 1977 The speech spectrum—some aspects of its significance for hearing aid selection and evaluation *Br. J. Audiol.* **11** 40–6

Caraway B J and Carhart R 1967 Influence of compressor action on speech intelligibility *J. Acoust. Soc. Am.* **41** 1424–33

Casselbrant M, Ingelstedt S and Ivarsson A 1977 Volume displacement of the tympanic membrane in the sitting position as a function of middle ear muscle activity *Acta Otolaryngol.* **84** 402–13

Causey G D and Beck L B 1974 Psychoacoustic calibration of the Telex 1470 earphone *J. Acoust. Soc. Am.* **55** 1088–9

Clarke F R and Bilger R C 1973 The theory of signal detectability and the measurement of hearing, in *Modern Developments in Audiology* ed J Jerger (New York: Academic Press) ch 12

Coles R R A and Priede V M 1975 Masking of the non-test ear in speech audiometry *J. Laryngol. Otol.* **89** 217–26

Colletti V 1975 Methodologic observations on tympanometry with regard to the probe tone frequency *Acta Otolaryngol.* **80** 54–60

—— 1976 Tympanometry from 200 to 2000 Hz probe tone *Audiology* **15** 106–19

Creten W L and van Camp K 1974 Transient and quasistatic tympanometry *Scand. Audiol.* **3** 39–42

Dadson R S and King J H 1952 A determination of the normal threshold of hearing and its relation to the standardisation of audiometers *J. Laryngol. Otol.* **66** 366–78

Dallos P J 1964 Dynamics of the acoustic reflex: phenomenological aspects *J. Acoust. Soc. Am.* **36** 2175–2183

—— 1973 Cochlear potentials and cochlear mechanics, in *Basic Mechanisms of Hearing* ed A Møller (New York: Academic Press) pp 335–76

—— 1975 Cochlear potentials, in *Human Communication and its Disorders* vol III ed D B Tower (New York: Raven) pp 69–80

Dallos P, Billone M, Durrant J D, Wang C Y and Raynor S 1972 Cochlear inner and outer hair cells: functional differences *Science* **177** 356–8

Dallos P and Wang C Y 1974 Bioelectric correlates of kanamycin intoxication *Audiology* **13** 277–89

Davis H 1961 Some principles of sensory receptor action *Physiol. Rev.* **41** 391–416

—— 1965 A model for the transducer action in the cochlea *Cold Spr. Harb. Symp. Quant. Biol.* **30** 181–90

—— 1976 Principles of electric response audiometry *Ann. Oto-Rhino-Laryngol.* **85** suppl. 28

Davis H, Hudgins C V *et al* 1946 The selection of hearing aids *Laryngoscope* **56** 85–115, 135–63

Davis H *et al* 1953 Acoustic trauma in the guinea pig *J. Acoust. Soc. Am.* **25** 1180–9

Davis H and Silverman S (eds) 1970 *Hearing and Deafness* (New York: Holt, Rinehart and Winston)

Davis H, Stevens S S *et al* 1947 *Hearing Aids: An Experimental Study of Design Objectives* (Cambridge, Mass.: Harvard University Press)

Deatherage B H, Eldredge D H and Davis H 1959 Latency of action potentials in the cochlea of the guinea pig *J. Acoust. Soc. Am.* **31** 479–86

Delany M E, Whittle L S, Cook J P and Scott V 1967 Performance studies on a new artificial ear *Acustica* **18** 231–7

Diem K and Lentner C (eds) 1970 *Documenta Geigy Scientific Tables* 7th edn (Macclesfield: Geigy Pharmaceuticals) pp 85–106

Dimmick F L and Olson R M 1941 The intensive difference limen in audition *J. Acoust. Soc. Am.* **12** 517–25

Djupesland G 1961 Recording intra-aural muscle reflexes by cutaneous stimulation in humans, preliminary report *Acta Otolaryngol.* **53** 397–404

—— 1962 Intra-aural muscular reflexes elicited by air current stimulation of the external ear *Acta Otolaryngol. Stockholm* **54** 143–53

—— 1964 Middle ear reflexes elicited by acoustic and non-acoustic stimulation *Acta Otolaryngol.* suppl. **188** 287–92

—— 1965 Electromyography of the tympanic muscles in man *Int. Audiol.* **4** 33–41

—— 1975 Advanced reflex considerations, in *Handbook of Clinical Impedance Audiometry* ed J Jerger (New York: American Electromedics Corporation Morgan Press) ch 5

Eggermont J J, Odenthal D W, Schmidt P H and Spoor A 1974 Electrocochleography: basic principles and clinical application *Acta Otolaryngol.* suppl. 316

Eliachar I and Northern J 1974 Studies in tympanometry: validation of the present technique for determining intra-tympanic pressures through the intact eardrum *Laryngoscope* **84** 247–55

Elliott L L 1975 Temporal and masking phenomena in persons with sensorineural hearing loss *Audiology* **14** 336–53

Elner A 1977 Quantitative studies of gas absorption from the normal middle ear *Acta Otolaryngol.* **83** 25–8

Evans E F 1975 Cochlear nerve and cochlear nucleus, in *Handbook of Sensory Physiology* ed W D Keidel and W D Neff vol V/2 Auditory Systems (Berlin: Springer) ch 1

—— 1976 The effective bandwidth of individual cochlear nerve fibres from pathological cochleas in the cat, in *Disorders of Auditory Function* vol 2 ed S D G Stephens (London: Academic Press) pl 115–26

Ewertsen H W (ed) 1971 Electroacoustic characteristics relevant to hearing aids *Scand. Audiol.* suppl. 1

Fechner G T 1860 *Elemente der Psychophysik* (English transl., H E Adler 1966 *Elements of Psychophysics* ed D H Howes and E G Boring (New York: Holt, Rinehart and Winston))

Feddersen W E, Sandel T T, Teas D C and Jeffress L A 1957 Localisation of high-frequency tones *J. Acoust. Soc. Am.* **29** 988–91

Feldman A S 1974 Eardrum abnormality and the measurement of middle ear function *Arch. Otolaryngol.* **99** 211–17

—— 1976 Tympanometry—procedures interpetations and variables, in *Acoustic Impedance and Admittance: The Measurement of Middle Ear Function* (Baltimore: Williams and Wilkins ch 6

Feldtkeller R and Zwicker E 1956 *Das Ohr als Nachrichtemempfänger* (Stuttgart: Hirzel)

Flanagan J L 1972 *Speech Analysis, Synthesis and Perception* (Berlin: Springer)

Flanagan J L and Guttmann N 1960 On the pitch of periodic pulses *J. Acoust. Soc. Am.* **32** 1308–19

Fleming D B and Rice C G 1969 New circuit development concepts in hearing aids *Int. Audiol.* **8** 517–23

Fletcher H 1953 *Speech and Hearing in Communication* (New York: van Nostrand)

Fletcher H and Munsen W A 1937 Relation between loudness and masking *J. Acoust. Soc. Am.* **9** 1–10

Flottorp G and Djupesland G 1970 Diphasic impedance change and its applicability to clinical work *Acta Otolaryngol.* suppl. **263** 200–4

Flottorp G, Djupesland G and Winther F 1971 The acoustic stapedius reflex in relation to critical bandwidth *J. Acoust. Soc. Am.* **49** 467–71

Fowler E P 1936 A method for the early detection of otosclerosis *Arch. Otolaryngol.* **24** 731–41

French N R and Steinberg J C 1947 Factors governing the intelligibility of speech sounds *J. Acoust. Soc. Am.* **19** 90–119

French St George M and Stephens S D G 1977 Acoustic reflex measurements of cochlear damage—a normative study *Br. J. Audiol.* **11** 111–19

Fry D B 1961 Word and sentence tests for use in speech audiometry *The Lancet* July 22 197–9

Greenwood D D 1961 Auditory masking and the critical band *J. Acoust. Soc. Am.* **33** 484–502

Grimaldi P 1976 The value of impedence testing in the diagnosis of middle ear effusion *J. Laryngol. Otol.* **90** 141–52

Groen J J 1956 The semicircular canal system and the organs of equilibrium *Phys. Med. Biol.* **1** 103–17, 225–42

Harker L A and van Wagoner R 1974 Application of impedance audiometry as a screening instrument *Acta Otolaryngol.* **77** 198–201

Hart C W and Naunton R F 1961 Frontal bone conduction tests in clinical audiometry *Laryngoscope* **71** 24–9

Haughton P M 1977 Validity of tympanometry for middle ear effusions *Arch. Otolaryngol.* **103** 505–13

Hawkins J E and Stevens S S 1950 The masking of pure tones and of speech by white noise *J. Acoust. Soc. Am.* **22** 6–13

von Helmholtz H 1863 *On the Sensations of Tone* (English transl. 1954) (New York: Dover)

Hirsh I J 1948 The influence of interaural phase on interaural summation and inhibition *J. Acoust. Soc. Am.* **20** 536–44

—— 1952 *The Measurement of Hearing* (New York: McGraw-Hill)

Hirsh I J, Davis H, Silverman S R, Reynolds E G, Eldert E and Benson R W 1952 Development of speech audiometry *J. Speech Hearing Dis.* **17** 321–37

Hirsh I J, Reynolds E G and Joseph M 1954 Intelligibility of different speech materials *J. Acoust Soc. Am.* **26** 530–8

Hodgson W R 1977 Speech acoustics and intelligibility, in *Hearing Aid Assessment and Use in Audiologic Habilitation* ed W R Hodgson and P H Skinner (Baltimore: Williams and Wilkins) ch 6

Holmquist J 1976 Auditory tubal function, in *Scientific Foundations of Otolaryngology* ed R Hinchcliffe and D Harrison (London: Heinemann) ch 17

Hood J D 1957 The principles and practice of bone conduction audiometry: a review of the present position *Proc. R. Soc. Med.* **50** 689–97

Hood J D and Poole J P 1977 Improving the reliability of speech audiometry *Br. J. Audiol.* **11** 93–102

Ingham J G 1958 Variations in cross-masking with frequency *J. Exp. Psychol.* **58** 199–205

Jeffress L A 1970 Masking, in *Modern Developments in Audiology* vol 1 ed J V Tobias (New York: Academic Press) ch 3

Jepsen O 1963 Middle-ear muscle reflexes in man, in *Modern Developments in Audiology* ed J Jerger (New York: Academic Press) ch 6

Jerger J 1960 Békésy audiometry in the analysis of auditory disorders *J. Speech Hearing Res.* **3** 275–87

—— 1970 Clinical experience with impedance audiometry *Arch. Otolaryngol.* **92** 311–24

—— 1972 Suggested nomenclature for impedance audiometry *Arch. Otolaryngol.* **96** 1–3

Jerger J, Anthony L, Jerger S and Mauldin L 1974a Studies in impedance audiometry III: Middle ear Disorders *Arch. Otolaryngol.* **99** 165–71

Jerger J, Burney P, Mauldin L and Crump B 1974b Predicting the hearing loss from the acoustic reflex *J. Speech Hearing Dis.* **39** 11–22

Jerger J, Jerger S and Mauldin L 1972 Studies in impedance audiometry I: Normal and sensorineural ears *Arch. Otolaryngol.* **96** 513–23

Jerger J and Thelin J 1968 Effects of electroacoustic characteristics of hearing aids on speech understanding *Bull. Prosthet. Res.* **11** 159–97

Jerger J and Tillman T 1960 A new method for the clinical determination of sensorineural acuity level (SAL) *Arch. Otolaryngol.* **71** 948–53

Jewett D L and Williston J S 1971 Auditory evoked far fields averaged from the scalp of humans *Brain* **94** 681–96

Johansson B 1973 The hearing aid as technical–audiological problem *Scand. Audiol.* suppl. **3** 55–76

Keith R 1977 An evaluation of predicting hearing loss from the acoustic reflex *Arch. Otolaryngol.* **103** 419–24

Kiang N Y S (ed) 1965 *Discharge Patterns of Single Fibers in the Cat's Auditory Nerve, Res. Monograph* 35 (Cambridge, Mass.: MIT Press)

—— 1975 *The Nervous System* vol 3: *Human Communication and Its Disorders* ed D B Tower (New York: Raven) pp 81–96

Kiang N Y S, Moxon E C and Levine R A 1970 Auditory nerve activity in cats with normal and abnormal cochleas, in *Sensorineural Hearing Loss* ed G E W Wolstenholme and J J Knight (London: Churchill) pp 241–68

Kim D and Molnar C 1975 *The Nervous System* vol 3: *Human Communication and Its Disorders* ed D B Tower (New York: Raven) pp 57–68

King R W P, Mimno H R and Wing A H 1945 *Transmission Lines, Antennas and Wave Guides* (New York: McGraw Hill)

Kinsler L E and Frey A R 1962 *Fundamentals of Acoustics* (New York: Wiley)

Kirikae I 1973 *Otolaryngology* vol 1: *Basic Sciences and Related Disciplines* ed D A Paperella and D A Shumrick (London: W B Saunders)

Klockhoff I 1961 Middle ear muscle reflexes in man *Acta Otolaryngol.* suppl. 164

Knight J J and Littler T S 1953 The technique of speech audiometry and a simple speech audimeter with masking generator for clincial use *J. Laryngol. Otol.* **67** 248–65

Koenig W, Dunn H K and Lacey L Y 1946 The sound spectrograph *J. Acoust. Soc. Am.* **18** 19–49

Kornhuber H H (ed) 1974 Vestibular system, in *Handbook of Sensory Physiology* vol VI parts 1 and 2 (Berlin: Springer)

Kryter K D 1962 Methods for the calculation and use of the articulation index *J. Acoust. Soc. Am.* **34** 1689–97

Langenbeck B 1931 Experimentelles und Theoretisches zue Frage der Horschwellenbestimmung *Pflüg. Arch. Ges. Physiol.* **226** 11–46

Licklider J C R 1946 Effects of amplitude distortion upon the intelligibility of speech *J. Acoust. Soc. Am.* **18** 429–34

Licklider J C R and Pollack I 1948 Effects of differentiation, integration and infinite peak clipping upon the intelligibility of speech *J. Acoust. Soc. Am.* **20** 42–51

Littler T S 1965 *The Physics of the Ear* (Oxford: Pergamon)

Littler T S, Knight J J and Strange P H 1952 Hearing by bone conduction and the use of bone conduction hearing aids *Proc. R. Soc. Med.* **45** 783–790

Lorente de Nó R 1976 Some unresolved problems concerning the cochlear nerve *Ann. Oto-Rhino-Laryngol.* **85** suppl. 34

Lybarger S F 1966 A discussion of hearing aid trends *Int. Audiol.* **5** 376–83

Lynn G and Carhart R 1963 Influence of attack and release in compression amplification on understanding of speech by hypoacusics *J. Speech Hearing Dis.* **28** 124–40

Lyregaard P E, Robinson D W and Hinchcliffe R 1976 *A Feasibility Study of Diagnostic Speech Audiometry* (NPL Acoustics Rep. Ac 73)

McMurray R F and Rudmose W 1956 An automatic audiometer for industrial medicine *Noise Control* **2** 33–6

Mawson S R 1974 *Diseases of the Ear* 3rd edn (London: Edward Arnold)

Mayer A M 1876 Researches in acoustics *Phil. Mag.* **2** 500–7

Medical Research Council 1947 *Hearing Aids and Audiometers, Spec. Rep. Ser. No* 261 (London: HMSO)

Metz O 1946 The acoustic impedance measured on normal and pathological ears *Acta Otolaryngol.* suppl. 63

—— 1952 Threshold of reflex contractions of muscles of middle ear and recruitment of loudness *Arch. Otolaryngol.* **55** 536–43

Michael P L and Bienvenue G R 1977 Real-ear threshold comparisons between the telephonics TDH-39 earphone with a metal outer shell and the TDH-39, TDH-49 and TDH-50 earphones with plastic outer shells *J. Acoust. Soc. Am.* **61** 1640–2

Miller R, Davies C and Gibson W 1976 Using the acoustic reflex threshold to predict the pure tone threshold *Br. J. Audiol.* **10** 51–4

Mills A W 1958 On the minimum audible angle *J. Acoust. Soc. Am.* **30** 237–46

—— 1972 Auditory localisation, in *Foundations of Modern Auditory Theory* vol II ed J V Tobias (New York: Academic Press) ch 8

Møller A R 1960 Improved technique for detailed measurement of middle ear impedance *J. Acoust. Soc. Am.* **32** 250–7

Munson W A and Wiener F M 1952 In search of the missing 6 dB *J. Acoust. Soc. Am.* **24** 498–501

Naunton R F 1963 The measurement of hearing by bone conduction, in *Modern Developments in Audiology* ed J Jerger (New York: Academic Press) ch 1

Niemeyer W and Sesterhenn G 1974 Calculating the hearing threshold from the stapedius reflex threshold for different sound stimuli *Audiology* **13** 421–7

Niemoeller A F, Silverman S R and Davis H 1970 Hearing aids, in *Hearing and Deafness* ed H Davis and S R Silverman (New York: Holt, Rinehart and Winston) ch 10

Nordmark J O 1970 Time and frequency analysis, in *Foundations of Modern Auditory Theory* vol I ed J V Tobias (New York: Academic Press) ch 2

Ohm G S 1843 Über die Definition des Tones *Ann. Phys. Chem.* **59** 513

Page C H and Vigoureux (eds) 1977 *The International System of Units* (London: HMSO)

Pascoe D P 1975 Frequency responses of hearing aids and their effects on the speech perception of hearing-impaired subjects *Ann. Oto-Rhino-Laryngol.* **84** suppl. 23

Pearson E S and Hartley H O (eds) 1966 *Biometrika Tables for Statisticians* (London: Cambridge University Press) pp 228-9

Pollack I 1949 Specification of sound pressure levels *Am. J. Psychol.* **62** 412-7

Potter R K, Knopp G A and Green H C 1947 *Visible Speech* (New York: Van Nostrand)

Priede V M and Coles R R A 1974 Interpretation of loudness recruitment tests—some new concepts and criteria *J. Laryngol. Otol.* **88** 641-62

Rainville M J 1955 Nouvelle methode d'assourdissement pour releve des courbes de conduction osseuse *J. Franç. Otolaryngol.* **4** 851-8

Ranke O F 1950 Theory of operation of the cochlea: a contribution to the hydrodynamics of the cochlea *J. Acoust. Soc. Am.* **22** 772-7

Rasmussen P E 1967 Middle ear and maxillary sinus during nitrous oxide anesthesia *Acta Otolaryngol.* **63** 7-16

Reneau J P and Hnatiow G Z 1975 *Evoked Response Audiometry* (Baltimore: University Park Press)

Renvall U and Holmquist J 1976 Tympanometry revealing middle ear pathology, in Recent Advances in Middle Ear Effusions *Ann. Oto-Rhino-Laryngol.* suppl. 25 209-15

Risberg A 1969 A critical review of work on speech analysing hearing aids *IEEE Trans. Audio Electroacoust.* **AU17** 290-7

Roberts T D M 1966 *Basic Ideas in Neurophysiology* (London: Butterworths)

—— 1976 Vestibular physiology, in *Scientific Foundations of Otolaryngology* ed R Hinchcliffe and D Harrison (London: Heinemann) ch 27

Robinson D W 1957 The subjective loudness scale *Acustica* **7** 217-33

—— 1971 A review of audiometry *Phys. Med. Biol.* **16** 1-24

Roworth D A A 1970 The re-coding of speech *Sound* **4** 95-105

Rudmose H W, Clark K C, Carlson F D, Eisenstein J C and Walker R A 1948 Voice measurements with an audio spectrometer *J. Acoust. Soc. Am.* **20** 503-12

Rutherford W 1886 A new theory of hearing *J. Anat. Physiol.* **21** 166-8

Salomon G and Starr A 1963 Electromyograph of middle ear muscles in man during motor activities *Acta Neurol. Scand.* **39** 161-8

Scharf B 1970 Critical bands, in *Foundations of Modern Auditory Theory* vol I ed J V Tobias (New York: Academic Press) ch 5

Scharf B and Hellman R P 1966 A model of loudness summation applied to impaired ears *J. Acoust. Soc. Am.* **40** 71-8

Schouten J F 1940 The residue, a new component in subjective sound analysis *Proc. K. Ned. Acad. Wet.* **43** 356-65

Schuster K 1934 Eine methode zum vergleich akustischer impedanzen *Phys. Z.* **35** 408-9

Seebeck A 1841 Beobachtungen über einige Bedungungen der Entstehung von Tönen *Ann. Physik* **53** ser. 2, 417-36

Seshadri S R 1971 *Fundamentals of Transmission Lines and Electromagnetic Fields* (Reading, Mass.: Addison-Wesley)

Sesterhenn G and Breuninger H 1977 Determination of hearing threshold for single frequencies from the acoustic reflex *Audiology* **16** 201–14

Shaw W A, Newman E B and Hirsh I J 1947 The difference between monaural and binaural thresholds *J. Exp. Psychol.* **37** 229–42

Shimada T and Lim D J 1971 The fiber arrangement of the human tympanic membrane *Ann. Oto-Rhino-Laryngol.* **80** 210–17

Shower E G and Biddulph R 1931 Differential pitch sensitivity of the ear *J. Acoust. Soc. Am.* **3** 275–87

Siebert W 1974 Ranke revisited—a simple short-wave cochlear model *J. Acoust. Soc. Am.* **56** 594–600

Sivian L J 1929 Speech power and its measurement *Bell Syst. Tech. J.* **8** 646–61

Sivian L J and White D S 1933 On the minimum audible field *J. Acoust. Soc. Am.* **4** 288–321

Small A M 1970 Periodicity pitch, in *Foundations of Modern Auditory Theory* vol I ed J V Tobias (New York: Academic Press) ch 1

Spoendlin H 1975 Neuroanatomical basis of cochlear coding mechanisms *Audiology* **14** 383–407

Steinberg J C, Montgomery H C and Gardner M B 1940 Results of the world's fair hearing tests *J. Acoust. Soc. Am.* **12** 291–301

Stephens S D G 1976 The input for a damaged cochlea—a brief review *Br. J. Audiol.* **10** 97–101

Stevens C F 1966 *Neurophysiology* (New York: Wiley)

Stevens S S 1935 The relation of pitch to intensity *J. Acoust. Soc. Am.* **6** 150–4

—— 1955 The measurement of loudness *J. Acoust. Soc. Am.* **27** 815–29

—— 1957 On the psychophysical law *Physiol. Rev.* **64** 153–81

—— 1961 The psychophysics of sensory functions, in *Sensory Communications* ed W A Rosenblith (Cambridge, Mass.: MIT Press)

Tanner W P and Sorkin R D 1972 The theory of signal detectability, in *Foundations of Modern Auditory Theory* vol II ed J V Tobias (New York: Academic Press) ch 2

Tobias J V and Zerlin S 1959 Lateralisation threshold as a function of stimulus duration *J. Acoust. Soc. Am.* **31** 1591–4

Terkildsen K and Nielsen S 1960 An electroacoustic impedance measuring bridge for clinical use *Arch. Otolaryngol.* **72** 339–46

Tonndorf J 1970 Cochlear mechanics and hydro-dynamics, in *Foundations of Modern Auditory Theory* vol I ed J V Tobias (New York: Academic Press) ch 6

Vanhuyse V J, Creten W L and van Camp K J 1975 On the W-notching of tympanograms *Scand. Audiol.* **4** 45

de Vries H L 1952 Brownian motion and the transmission of energy in the cochlea *J. Acoust. Soc. Am.* **24** 527–33

Ward W D 1970 Musical perception, in *Foundations of Modern Auditory Theory* vol I ed J V Tobias (New York: Academic Press) ch 11

Weber E H 1834 *De pulsu, Resorptione, Auditu et Tactu: Annotationes Anatomica et Pysiologicae* (Leipzig: Köhler) p 175

Webster F A 1951 The influence of interaural phase on masked thresholds *J. Acoust. Soc. Am.* **23** 452–62

Wegel R L and Lane C E 1924 The auditory masking of one pure tone by another and its probable relation to the dynamics of the inner ear *Phys. Rev.* **23** 266–85

Weiss E 1960 An air damped artificial mastoid *J. Acoust. Soc. Am.* **32** 1582–8

Wever E G 1949 *Theory of Hearing* (New York: Wiley)

Wever E G and Bray C W 1930 Action currents in the auditory nerve in response to acoustical stimulation *Proc. Nat. Acad. Sci. Wash.* **16** 344–50

—— 1937 The perception of low tones and the resonance-volley theory *J. Psychol.* **3** 101–14

Wever E G, Bray C W and Lawrence M 1940 A quantitative study of combination tones *J. Exp. Psychol.* **27** 469–96

Wever E G and Lawrence M 1954 *Physiological Acoustics* (Princeton, NJ: Princeton University Press)

Wever E G and Wedell C H 1941 Pitch discrimination at high frequencies *Psychol. Bull.* **38** 727

Wheeler L J and Dickson E D 1952 The determination of the threshold of hearing *J. Laryngol. Otol.* **66** 379–95

Whittle L S 1965 A determination of the normal threshold of hearing by bone conduction *J. Sound Vibration* **2** 227–48

Wilber L A 1972 Comparability of two commercially available artificial mastoids *J. Acoust. Soc. Am.* **52** 1265–6

Woodworth R S 1938 *Experimental Psychology* (New York: Holt) p 521

Yoshie N and Ohashi T 1969 Clinical use of cochlear nerve action potential in man for differential diagnosis of hearing losses *Acta Otolaryngol.* suppl. 252 71–87

Zwicker E 1958 Über phychologische und methodische Grundlagen der Lautheir *Acustica* **8** 237–58

Zwislocki J 1950 Theory of the acoustical action of the cochlea *J. Acoust. Soc. Am.* **22** 778–84

—— 1963 An acoustic method for clinical examination of the ear *J. Speech Hearing Res.* **6** 303–14

—— 1965 Analysis of some auditory characteristics, in *Handbook of Mathematical Psychology* ed R D Luce *et al* (New York: Wiley)

—— 1975 *The Nervous System* vol 3: *Human Communication and its Disorders* ed D B Tower (New York: Raven) pp 45–55

Zwislocki J and Feldman R S 1956 Just noticeable differences in dichotic phase *J. Acoust. Soc. Am.* **28** 860–4

—— 1970 *Acoustic Impedance of Pathological Ears, ASHA Monograph* 15 pp1–42

Index